Collins
INTERNATIONAL
LOWER SECONDARY

Computing
Student's Book
7

Rebecca Franks, Dr Tracy Gardner and Liz Smart

William Collins' dream of knowledge for all began with the publication of his first book in 1819.

A self-educated mill worker, he not only enriched millions of lives, but also founded a flourishing publishing house. Today, staying true to this spirit, Collins books are packed with inspiration, innovation and practical expertise.

They place you at the centre of a world of possibility and give you exactly what you need to explore it.

Collins. Freedom to teach.

Published by Collins

An imprint of HarperCollins*Publishers*
The News Building, 1 London Bridge Street, London,
SE1 9GF, UK

HarperCollins*Publishers*
Macken House, 39/40 Mayor Street Upper, Dublin 1,
D01 C9W8, Ireland

Browse the complete Collins catalogue at
collins.co.uk

© HarperCollins*Publishers* Limited 2024

10 9 8 7 6 5 4 3 2 1

ISBN 978-0-00-868402-0

All rights reserved. No part of this publication may be reproduced, stored in a retrieval system, or transmitted in any form by any means, electronic, mechanical, photocopying, recording or otherwise, without the prior written permission of the Publisher or a licence permitting restricted copying in the United Kingdom issued by the Copyright Licensing Agency Ltd, 5th Floor, Shackleton House, 4 Battle Bridge Lane, London SE1 2HX.

British Library Cataloguing-in-Publication Data

A catalogue record for this publication is available from the British Library.

Authors: Rebecca Franks, Dr Tracy Gardner and Liz Smart
Publisher: Catherine Martin
Product manager: Saaleh Patel
Project manager: Just Content Ltd
Development editor: Julie Bond
Copy editor: Becca Law
Proofreader: Laura Connell and Tim Jackson
Cover designer: Gordon McGilp
Cover image: Amparo Barrera, Kneath Associates
Internal designer: Steve Evans, Planet Life Art
Illustration: Jouve India Ltd
Typesetter: Ken Vail Graphic Design
Production controller: Lyndsey Rogers
Printed and bound by Martins the Printers

This book contains FSC™ certified paper and other controlled sources to ensure responsible forest management.

For more information visit: harpercollins.co.uk/green

Contents

Introduction: How to use this book	**v**

Chapter 1: Our digital world — 1

1.1	Safe and healthy use of technology	2	**1.5**	Make your own machine learning project	16
1.2	Responsible participation in online communities	6	**1.6**	Share findings on AI and machine learning	20
1.3	Applications of AI	8			
1.4	Train a machine learning model	11			

Chapter 2: Content creation — 22

2.1	Collaborative working	23	**2.4**	Plan a news report	32
2.2	Usage rights	26	**2.5**	Record a news report	34
2.3	Future technologist reliability	30	**2.6**	Broadcast a news report	36

Chapter 3: Create with code 1 — 38

3.1	Logic gates	39	**3.5**	Build a program from a flowchart	47
3.2	Selection in flowcharts	42	**3.6**	Showcase a flowchart and a program	48
3.3	Debug flowcharts	44			
3.4	Design a flowchart	45			

Chapter 4: How computers work — 50

4.1	Software and automation	51	**4.4**	Design your pixel art	61
4.2	Binary representation	55	**4.5**	Create your pixel art	62
4.3	Represent images in binary	58	**4.6**	Showcase your artwork	63

Chapter 5: Create with code 2 — 65

5.1	Python: input and output	66	**5.5**	Complete your recommendation project	81
5.2	Python: operators, conditions and data types	71	**5.6**	Share your recommendation project	83
5.3	Python: Boolean logic, comments and functions	75			
5.4	Start your recommendation project	79			

Chapter 6: Connect the world — 85

6.1	Internet standards: DNS and HTTPS	86	**6.5**	Complete your encryption algorithm — 99
6.2	Wireless data transmission	90	**6.6**	Cracking the code — 101
6.3	Encryption and data security	93		
6.4	Create your own encryption algorithm	97		

Chapter 7: The power of data — 103

7.1	Modelling data to make decisions	104	**7.4**	Ask your audience	118
7.2	Organising and formatting data	110	**7.5**	Analyse your data	119
7.3	Search and capture data	115	**7.6**	Present your decision	120

Chapter 8: Create with code 3 — 122

8.1	Project plans	123	**8.4**	Create your device	131
8.2	Test plans	127	**8.5**	Test and refine your device	132
8.3	Prototypes	129	**8.6**	Showcase your device	133

Glossary of key terms — 135

Acknowledgements — 137

Introduction: How to use this book

The Collins Stage 7 Student's Book and Workbook offer a rich programme of skills development, based on a varied and stimulating set of projects grounded in real-world contexts.

The series is built around six themes relating to computing, ICT and digital literacy. These are:
- Our digital world – Providing the tools to safely navigate the digital world around us
- Content creation – Creating content using a variety of software, from office tools to video production
- Create with code – Exploring the fundamentals of programming and computational thinking skills
- How computers work – Lifting the lid on the specific technologies that make up a computer system
- Connect the world – Exploring how the world is connected through networks, the internet and the World Wide Web
- The power of data – Essential skills in collecting, analysing and presenting data linked to real-world activities that create or empower change.

Each chapter is carefully organised to develop essential knowledge and skills whilst working towards the creation of a final project. The final projects are designed to boost your creative skills and to give you the opportunity to make decisions and develop outputs that matter to you. All chapters end with a showcase lesson that gives you the opportunity to develop your presenting skills, whilst gaining valuable feedback on your work.

Stage 7 has eight chapters that relate directly to these six themes, with three chapters dedicated to creating with code.

You will delve into the safe and respectful utilisation of online collaboration tools. Engaging practical activities guide you through the exploration of new machine learning and artificial intelligence concepts. Enhance your software development skills with effective use of flowcharts and prototyping emphasised. Binary representation of data is studied, leading to the collaborative creation of an art installation. Technical aspects of networking, including encryption, are also explored.

Key features of the Student's Book

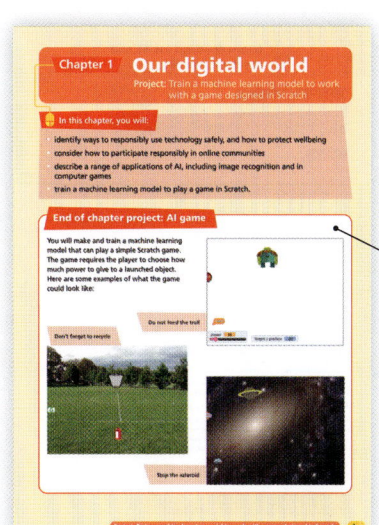

The opening page of each chapter provides an overview of the key activities as well as an explanation of the final project, with examples.

Each lesson begins with a helpful reminder of relevant concepts you have already learned about.

'Discuss' exercises encourage you to develop a shared understanding and gain inspiration from your classmates. Other types of activities include Build and Investigate.

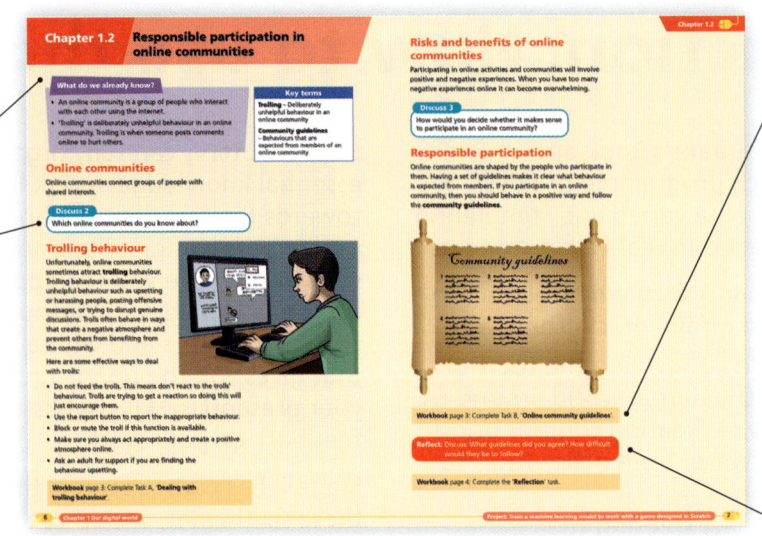

Clear references are provided to tasks to complete in the Workbook.

Reflection exercises encourage you to think about what you have learned and achieved and how you feel about your progress.

'Stay safe' tips remind you about potential hazards in the digital world.

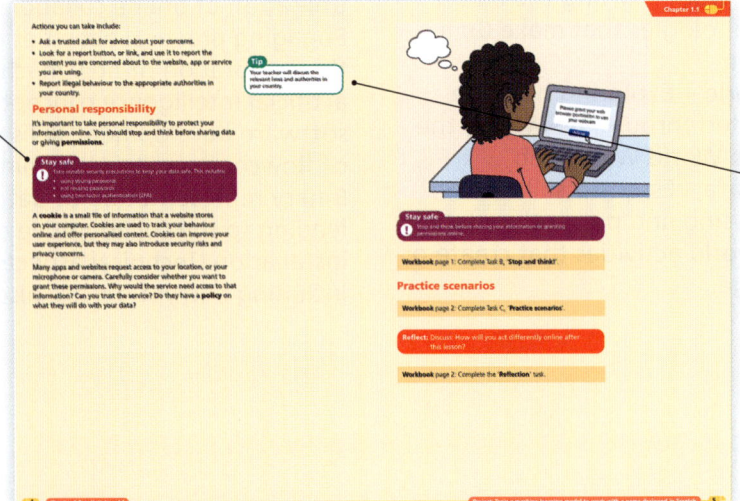

Helpful tips from the authors support you to avoid common problems and make your work the best it can be.

Showcasing your work develops your communication skills and enables you to receive feedback from others.

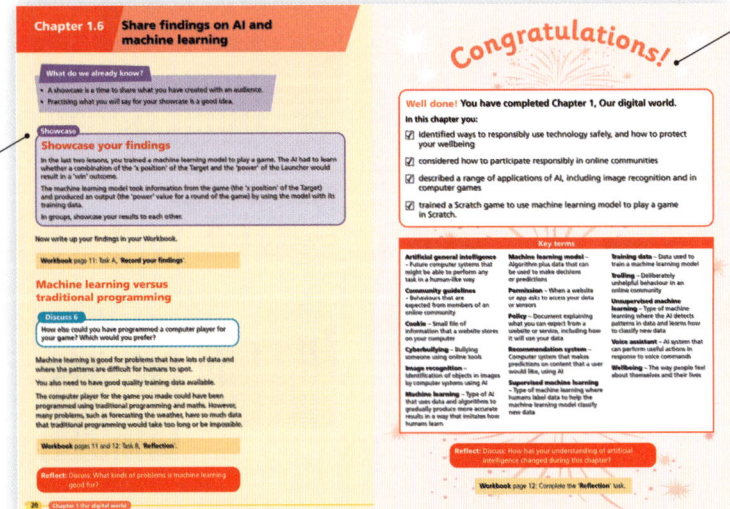

Celebrate what you have achieved and recap your learning at the end of each chapter.

Introduction: How to use this book

Chapter 1: Our digital world

Project: Train a machine learning model to work with a game designed in Scratch

In this chapter, you will:

- identify ways to responsibly use technology safely, and how to protect wellbeing
- consider how to participate responsibly in online communities
- describe a range of applications of AI, including image recognition and in computer games
- train a machine learning model to play a game in Scratch.

End of chapter project: AI game

You will make and train a machine learning model that can play a simple Scratch game. The game requires the player to choose how much power to give to a launched object. Here are some examples of what the game could look like:

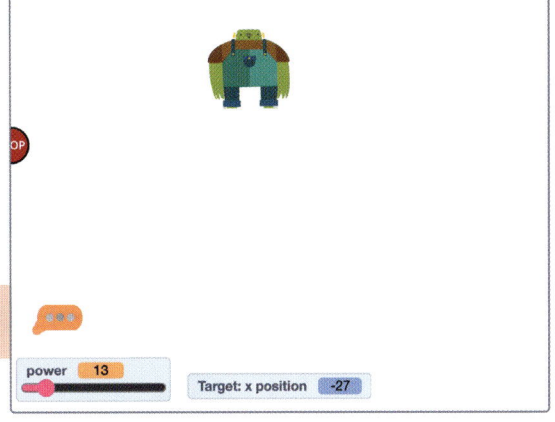

Do not feed the troll

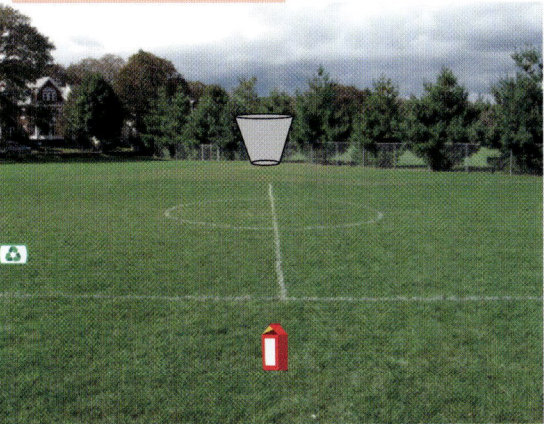

Don't forget to recycle

Stop the asteroid

Chapter 1.1 Safe and healthy use of technology

What do we already know?

- A digital footprint is the trail of information left behind by your activity in the digital world.
- You should consider your wellbeing when using online tools.
- There are formal procedures for reporting illegal and offensive online behaviour, including cyberbullying.

Key terms

Wellbeing – The way people feel about themselves and their lives

Cyberbullying – Bullying someone using online tools

Permission – When a website or app asks to access your data or sensors like camera or microphone

Cookie – Small file of information that a website stores on your computer

Policy – Document explaining what you can expect from a website or service, including how it will use your data

Your digital footprint

Your digital footprint could include:

- profile pages for computer gaming accounts
- comments and likes on online content
- photos you have shared
- questions you have asked or answered
- messages in group chats
- a portfolio of your artwork.

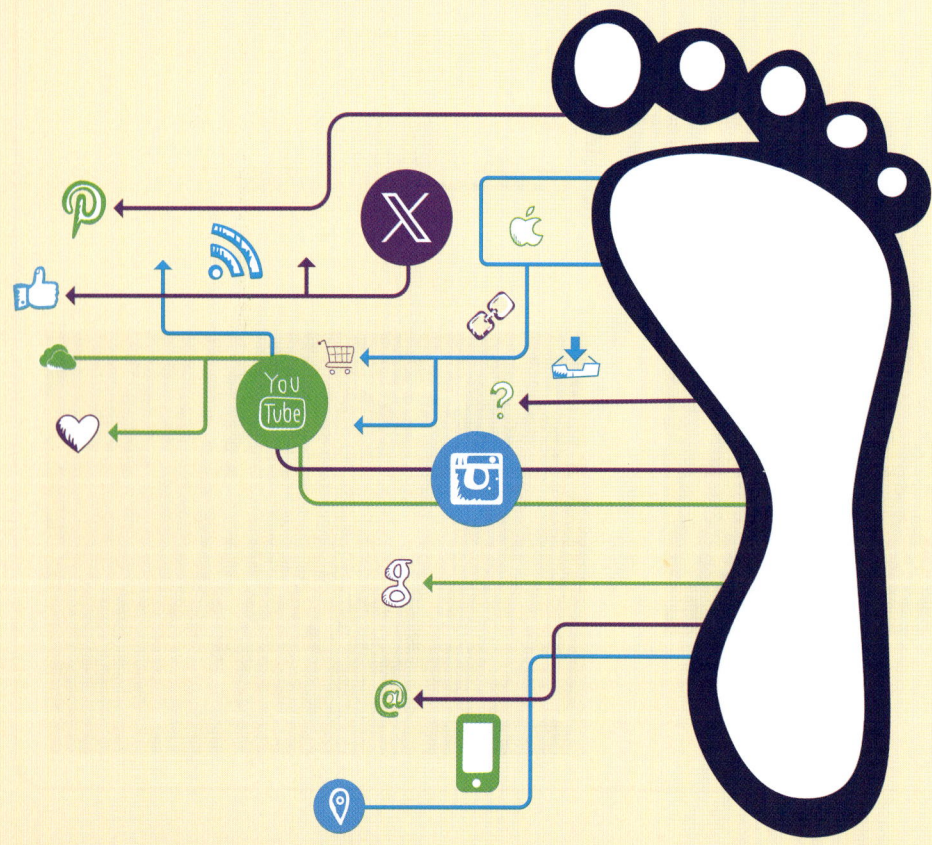

DIGITAL FOOTPRINT

Chapter 1.1

> **Discuss 1**
>
> What are the positive and negative consequences of having a digital footprint?

Protecting your wellbeing online

Participating in online activity can be a positive experience, but there can also be negative consequences for your **wellbeing**.

Wellbeing is the way people feel about themselves and their lives. It refers to overall health, including physical, mental and emotional health, and social relationships.

Tips for protecting your wellbeing online:

- Check your privacy settings to make sure you are not sharing information you do not want others to see.
- Limit your social media time to an amount that has a positive impact on your wellbeing.
- Leave online groups that are not useful to you.
- Discuss any concerns with supportive adults.
- Make sure your contributions are positive and won't have a negative impact on others.

> **Workbook** page 1: Complete Task A, '**Protecting your wellbeing online**'.

Taking action

Even if you are very careful, you may sometimes see behaviour or content online that is worrying, offensive or even illegal.

Bullying someone using online tools is called **cyberbullying**. In many countries, cyberbullying is illegal. There are also laws relating to activities such as online fraud and child exploitation.

It's important to know what to do to protect yourself and others when you come across inappropriate content.

Project: Train a machine learning model to work with a game designed in Scratch

Actions you can take include:

- Ask a trusted adult for advice about your concerns.
- Look for a report button, or link, and use it to report the content you are concerned about to the website, app or service you are using.
- Report illegal behaviour to the appropriate authorities in your country.

> **Tip**
> Your teacher will discuss the relevant laws and authorities in your country.

Personal responsibility

It's important to take personal responsibility to protect your information online. You should stop and think before sharing data or giving **permissions**.

> **Stay safe**
>
> Take sensible security precautions to keep your data safe. This includes:
> - using strong passwords
> - not reusing passwords
> - using two-factor authentication (2FA).

A **cookie** is a small file of information that a website stores on your computer. Cookies are used to track your behaviour online and offer personalised content. Cookies can improve your user experience, but they may also introduce security risks and privacy concerns.

Many apps and websites request access to your location, or your microphone or camera. Carefully consider whether you want to grant these permissions. Why would the service need access to that information? Can you trust the service? Do they have a **policy** on what they will do with your data?

Chapter 1.1

Stay safe
 Stop and think before sharing your information or granting permissions online.

Workbook page 1: Complete Task B, '**Stop and think!**'.

Practice scenarios

Workbook page 2: Complete Task C, '**Practice scenarios**'.

Reflect: Discuss: How will you act differently online after this lesson?

Workbook page 2: Complete the '**Reflection**' task.

Project: Train a machine learning model to work with a game designed in Scratch

Chapter 1.2 Responsible participation in online communities

> **What do we already know?**
> - An online community is a group of people who interact with each other using the internet.
> - 'Trolling' is deliberately unhelpful behaviour in an online community. Trolling is when someone posts comments online to hurt others.

> **Key terms**
> **Trolling** – Deliberately unhelpful behaviour in an online community
> **Community guidelines** – Behaviours that are expected from members of an online community

Online communities

Online communities connect groups of people with shared interests.

> **Discuss 2**
> Which online communities do you know about?

Trolling behaviour

Unfortunately, online communities sometimes attract **trolling** behaviour. Trolling behaviour is deliberately unhelpful behaviour such as upsetting or harassing people, posting offensive messages, or trying to disrupt genuine discussions. Trolls often behave in ways that create a negative atmosphere and prevent others from benefiting from the community.

Here are some effective ways to deal with trolls:

- Do not feed the trolls. This means don't react to the trolls' behaviour. Trolls are trying to get a reaction so doing this will just encourage them.
- Use the report button to report the inappropriate behaviour.
- Block or mute the troll if this function is available.
- Make sure you always act appropriately and create a positive atmosphere online.
- Ask an adult for support if you are finding the behaviour upsetting.

Workbook page 3: Complete Task A, '**Dealing with trolling behaviour**'.

Risks and benefits of online communities

Participating in online activities and communities will involve positive and negative experiences. When you have too many negative experiences online it can become overwhelming.

> **Discuss 3**
> How would you decide whether it makes sense to participate in an online community?

Responsible participation

Online communities are shaped by the people who participate in them. Having a set of guidelines makes it clear what behaviour is expected from members. If you participate in an online community, then you should behave in a positive way and follow the **community guidelines**.

Workbook page 3: Complete Task B, '**Online community guidelines**'.

Reflect: Discuss: What guidelines did you agree? How difficult would they be to follow?

Workbook page 4: Complete the '**Reflection**' task.

Chapter 1.3 Applications of AI

What do we already know?

- Artificial intelligence (AI) is a simulation of human intelligence within computer systems.
- AI is used within common productivity software such as predictive text.
- Trolling is deliberately unhelpful behaviour in an online community.
- Scratch is a program that allows you to create your own apps.

Key terms

Artificial general intelligence – Future computer systems that might be able to perform any task in a human-like way

Voice assistant – AI system that can perform useful actions in response to voice commands

Recommendation system – Computer system that makes predictions on content that a user would like, using AI

Image recognition – Identification of objects in images by computer systems using AI

What is AI?

Artificial intelligence (AI) is a simulation of human intelligence within computer systems.

Figure 1.1 Using artificial intelligence

AI usually refers to computer systems that have been designed to perform a specific set of tasks.

The term **artificial general intelligence** describes future computer systems that might be able to perform any task in a human-like way.

Discuss 4

Some people use the term artificial intelligence when they mean *artificial general intelligence*. What is the difference?

Figure 1.2 Artificial general intelligence

Chapter 1 Our digital world

Applications of AI

AI is used in many computer systems in modern life. You might have experienced the use of AI in a **voice assistant** that can answer questions or play music for you.

1. Voice assistant:

2. Recommendation system:

3. Facial recognition:

4. Self-driving car:

AI is used to make recommendations on entertainment platforms. This is known as a **recommendation system**. AI is also used in games to make the characters you play against act in a smart way and challenge you, or to make them show appropriate emotions and behaviour. Some robots and machines use AI to help them to do things such as cleaning your house or even driving cars.

> Workbook page 5: Complete Task A, '**AI definitions**'.

AI and image recognition

AI is widely used for **image recognition**. AI image recognition systems can accurately identify and classify objects in images, enabling various applications including:

- facial recognition for computer login or auto tagging people on social media
- identifying physical objects such as plants or products
- self-driving cars recognising obstacles
- analysing medical images.

AI and computer games

AI is increasingly being used in the video games industry for many features including:

- controlling non-player characters
- generating behaviour and dialogue
- generating realistic landscapes and music
- adapting difficulty and balancing teams.

Workbook page 5: Complete Task B, '**AI in your life**'.

Learning to play a computer game

In the next lesson, you will create a project with AI that can play a computer game. First, you need to play the game yourself to understand how it works.

Investigate 1

Your teacher will provide a link to the game.

In 'Do not feed the troll', you need to prevent replies being sent to the troll because it's much better to ignore a troll. You need to stop the messages by pressing the space bar at the right time.

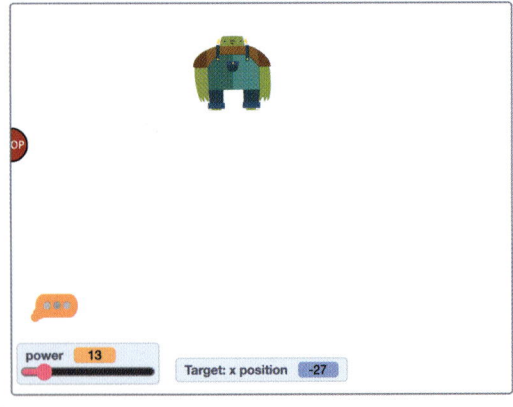

Some AI systems imitate human-like learning.

Workbook page 6: Complete Task C, '**How do you learn to play a computer game?**'.

Reflect: Discuss: How did you learn to play the game?

Workbook page 6: Complete the '**Reflection**' task.

Chapter 1.4 Train a machine learning model

What do we already know?

- An algorithm is a sequence of instructions to complete a task or solve a problem.
- (Traditional) computer programming is computer programming with step-by-step instructions.

Key terms

Machine learning – Type of AI that uses data and algorithms to gradually produce more accurate results in a way that imitates how humans learn

Machine learning model – Algorithm plus data that can be used to make decisions or predictions

Training data – Data used to train a machine learning model

Supervised machine learning – Type of machine learning where humans label data to help the machine learning model to classify new data

Unsupervised machine learning – Type of machine learning where the AI detects patterns in data and learns how to classify new data

What is machine learning?

Machine learning is a kind of AI that uses data and algorithms to gradually produce more accurate results in a way that imitates how humans learn.

Machine learning is used in many applications, including chat bots, recommendation systems, image recognition and computer games.

A **machine learning model** uses an algorithm and data to make decisions or predictions.

Training a machine learning model

In machine learning, you train a machine learning model using **training data**. Once you have a trained model, it can be used to produce outputs for new inputs.

In **supervised machine learning**, a human labels data so that the machine learning model has examples that match inputs to outputs. For example, you can train a machine learning model to recognise objects in images. This is done by providing lots of example images that have been labelled by humans.

In **unsupervised machine learning**, the machine learning algorithm has to discover the relationship between inputs and outputs.

Workbook page 7: Task A, '**Complete the AI definitions**'.

Workbook page 7: Task B, '**Labelling data**'.

Figure 1.3 If you've ever filled in a Completely Automated Public Turing test to tell Computers and Humans Apart (CAPTCHA) then you may have been labelling data for a supervised machine learning model.

Project: Train a machine learning model to work with a game designed in Scratch

Train a machine learning model

You are going to use supervised machine learning to train a machine learning model to play 'Do not feed the troll'. You will train the machine learning model to play the game by collecting examples of 'win' and 'lose' game outcomes.

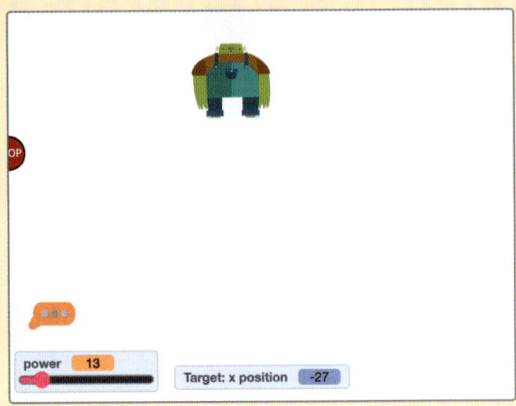

Discuss 5

What data do you need to decide whether an attempt will be successful?

Set up the machine learning model

Build 1:

Use the link provided by your teacher to open the website.

1. Click 'Get started'
2. Click 'Try it now'
3. Click 'Add a new project'
4. Set Project Name: Do not feed the troll
5. Set Project Type: recognising numbers
6. Add two values: target x (number) and power (number)
7. Set where you want to store your project (web browser)
8. Click 'Create'
9. Click on your project
10. Click 'Train'
11. Click 'Add new label' to add a label called 'win' and another called 'lose'

You have created labels for a machine learning model for supervised learning.

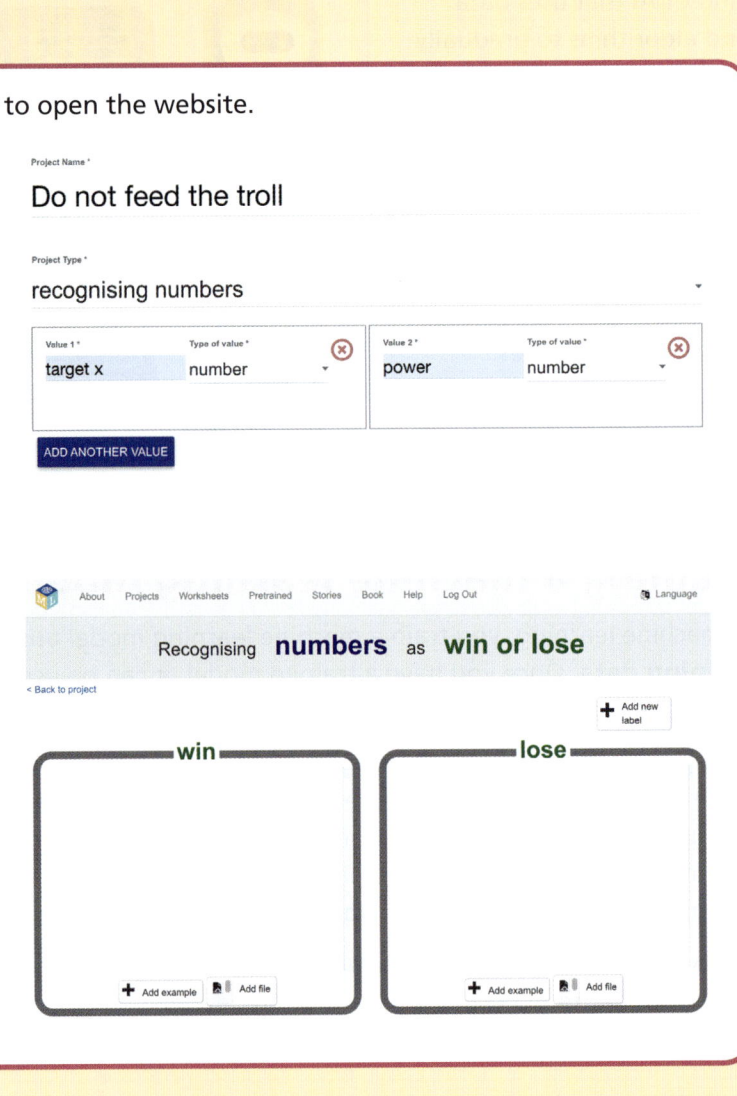

Gather training data

Chapter 1.4

Build 2:

1. Click 'Back to project'.
2. Click 'Make' then 'Scratch 3' then 'straight into Scratch' (you are going to use Scratch to add training data). Scratch will open in a new browser tab with some additional blocks.
3. Switch Scratch to use high contrast blocks to match the colours in this book:
4. In Scratch, choose 'File' then 'Load from your computer' then choose the 'Do not feed the troll' project. Your teacher will tell you how to access the project.

5. Find the scripts for 'win' and 'lose' on the Target sprite – these scripts run when the game has those outcomes.
6. Click on the 'Do not feed the troll' blocks category.
7. Add a 'add training data target x' block to both scripts to tell the machine learning model when a win or lose outcome occurs. Set the drop-down in the lose script to 'lose'.
8. Add the 'x position' and 'power' variables so that the machine learning model knows which input data lead to the outcome.

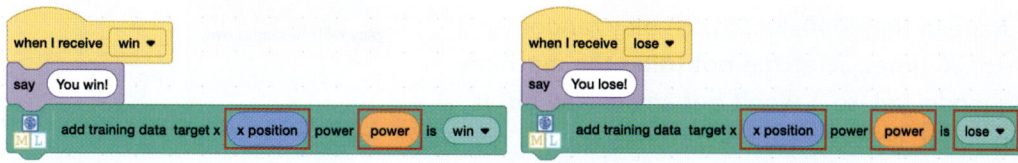

9. Play a round of the game by clicking the green flag and then pressing the space key to apply power to the Launcher sprite. Don't worry whether you win or lose.
10. Switch to your 'Machine Learning for Kids' browser tab and click 'Back to Project' and then 'Train'.

You should see the result of your first round of playing the game.

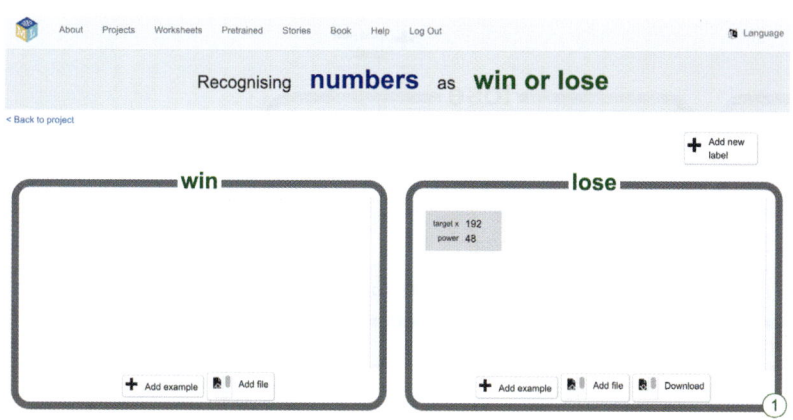

Record your results by completing

Workbook page 7: Task C, **'First training data'**.

Project: Train a machine learning model to work with a game designed in Scratch

Build 3:

1. Go back to Scratch and play about five more times. Try to get a mix of win and lose.

2. Go back to the 'Machine Learning for Kids' browser tab and view the training data. Click 'Refresh' ↻ in your web browser to see the new data.

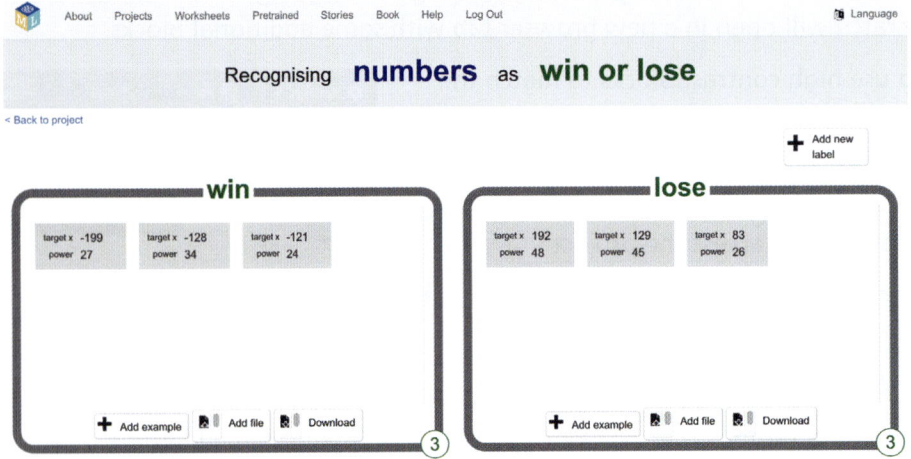

3. Return to Scratch and click on the 'Launcher' sprite.

4. Find the 'When flag clicked' script and change the 'player chooses power' to 'play with random power'.

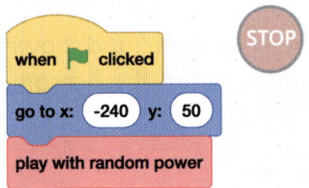

5. Click the green flag to make Scratch play randomly. Repeat for 20 times. Scratch is not using the machine learning model yet, so it won't get any better, but it will help you to collect a variety of data.

6. Switch to the 'Machine Learning for Kids' browser tab and refresh the Training data to see the attempts. Your numbers will be different to these.

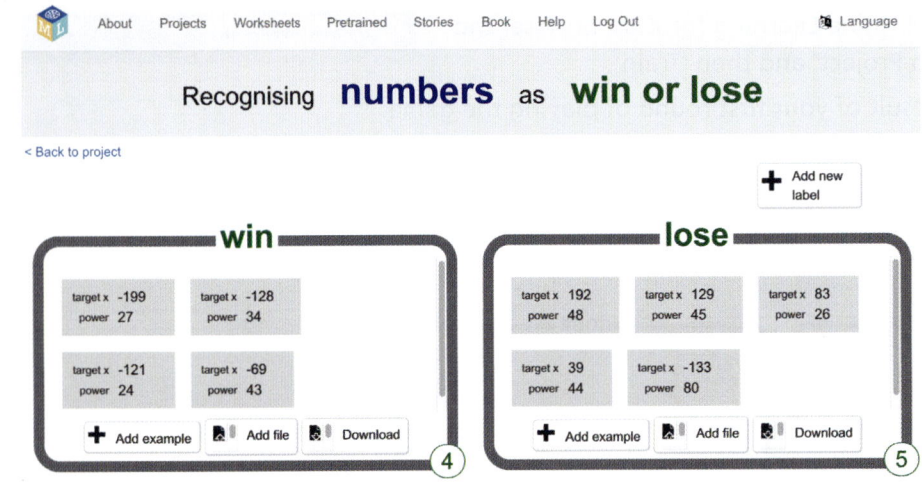

Record your results by completing

Workbook page 8: Task D, '**After 20 attempts**'.

You have provided training data for your machine learning model.

14   Chapter 1 Our digital world

Chapter 1.4

Play using machine learning

Build 4:

1. Update the Launcher sprite 'When flag clicked' script to play using machine learning.

   Make sure your script looks like this:

   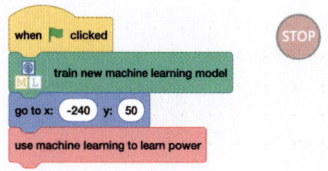

2. Find the 'define [use machine learning to learn power]' script. It generates random power values but checks them with the machine learning model before using them. It will only use a value if the machine learning model predicts that it will lead to a win outcome, or if the machine learning model is taking too many attempts to get a good result (machine learning can use a lot of computing resources).

   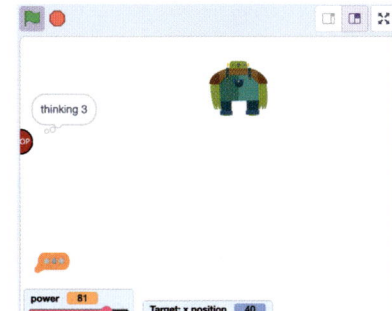

3. Update the 'define [use machine learning to learn power]' script to check the power value with the machine learning model before using it.

4. Click the green flag and observe the results.

   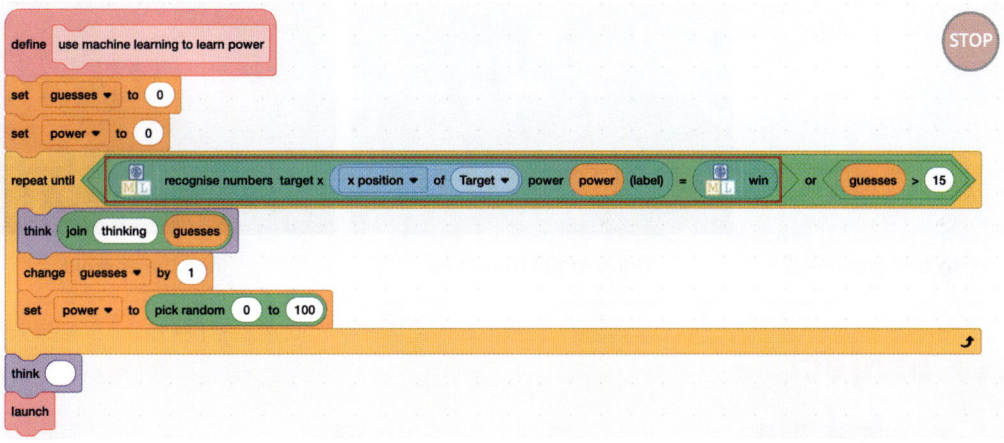

5. Play many times (at least 20) and watch how the behaviour of the AI changes.

Record your results by completing

Workbook page 8: Task E, '**After many attempts**'.

The machine learning model will keep learning from its attempts, so it will keep getting better. Check that new training data has been added.

Reflect: Discuss: How did the training data improve the AI?

Tip
Note that your training data will not be saved if you don't have an account. That's okay as you will not need this data for future lessons.

Workbook page 8: Complete the '**Reflection**' task.

Project: Train a machine learning model to work with a game designed in Scratch

Chapter 1.5 Make your own machine learning project

What do we already know?

- You can train a machine learning model by playing a game.
- The machine learning model uses training data to play the game.

Project brief

Train your own machine learning project

In this lesson, you will personalise the 'Do not feed the troll' project from the previous lesson. In your project, the player will launch an object to intersect with a projectile.

You will then train a machine learning model to play your project.

Here are some examples of the games or simulations that you could make:

Do not feed the troll (modified)

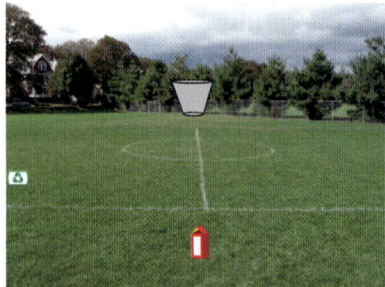
Don't forget to recycle

Stop the asteroid

Plan your project

Workbook page 9: Task A, '**Plan your project**'.

to choose costumes for the 'Target', 'Launcher' and 'Projectile' sprites.

- Choose the costume(s) for the 'Target' sprite. (You could just change Frank's costume.)
- Choose the costume for the 'Launcher' sprite, which will have its speed controlled by the 'power' variable.
- Choose the costume for the 'Projectile' sprite, which the Launcher needs to intercept to stop it reaching the target.
- Change the sizes of the 'Launcher' and 'Projectile' sprite to adjust the difficulty.
- Add a backdrop.

Personalise your project

Build 5:

You will make your game in regular Scratch before using it with Machine Learning for Kids.

1. Your teacher will tell you how to open the starter project.
2. Personalise your project by adding the costumes and backdrop (if any) from your plan.
3. Change the 'when flag clicked' script on the 'Target' sprite to use the costume(s) you have added.
4. Adjust the size of the 'Launcher' and 'Projectile' sprites until you are happy with the difficulty level.
5. Choose 'File' then 'Save to your computer' to save the project.
6. When you are sure you have saved your project to your computer, close Scratch.

Set up your machine learning model

> **Tip**
> Look back at the detailed instructions from last lesson if you need more help.

Build 6:

You teacher will provide you with a link.

1. Create a new project with the name of your project and add the same two number values for the data:

2. Add win and lose labels.

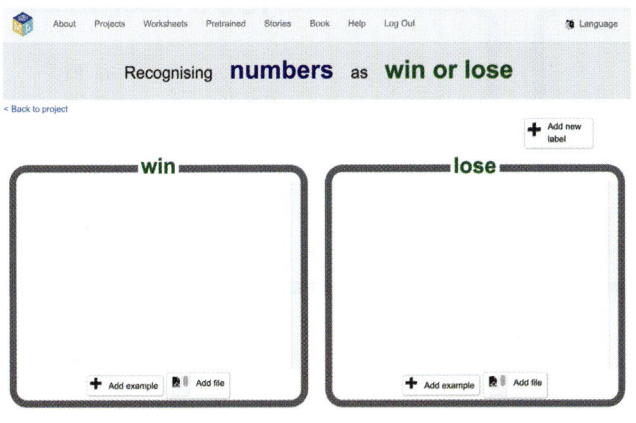

Project: Train a machine learning model to work with a game designed in Scratch

Train your machine learning model

Build 7:

1. Click 'Back to project'. Click 'Make'. Click 'Scratch 3'. Click 'straight into Scratch'.
2. Load your personalised Scratch project using 'File' then 'Load from your computer'.
3. Add code blocks to the win and lose scripts on the 'Target' sprite.

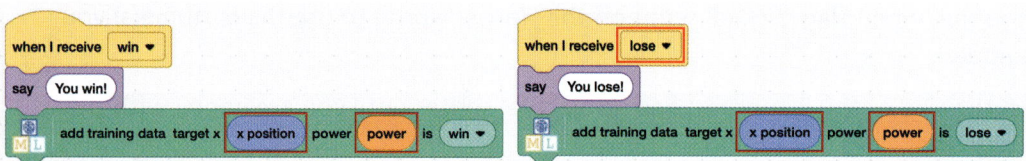

4. Train your machine learning model by playing yourself, playing randomly or a combination. On the launcher sprite, either:

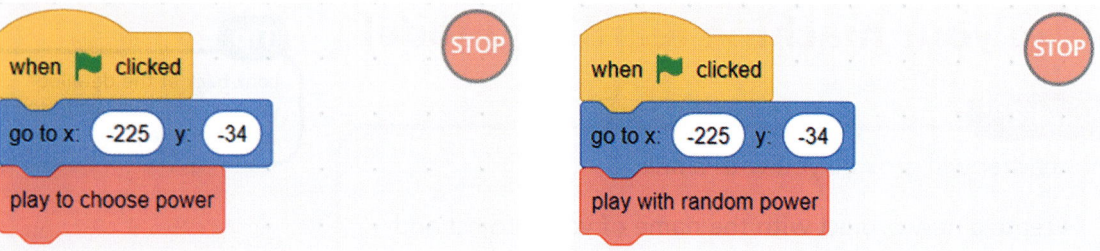

5. Check that you can see the training data. Click 'Refresh' in your web browser ↻ to see the latest data. Your numbers will be different.

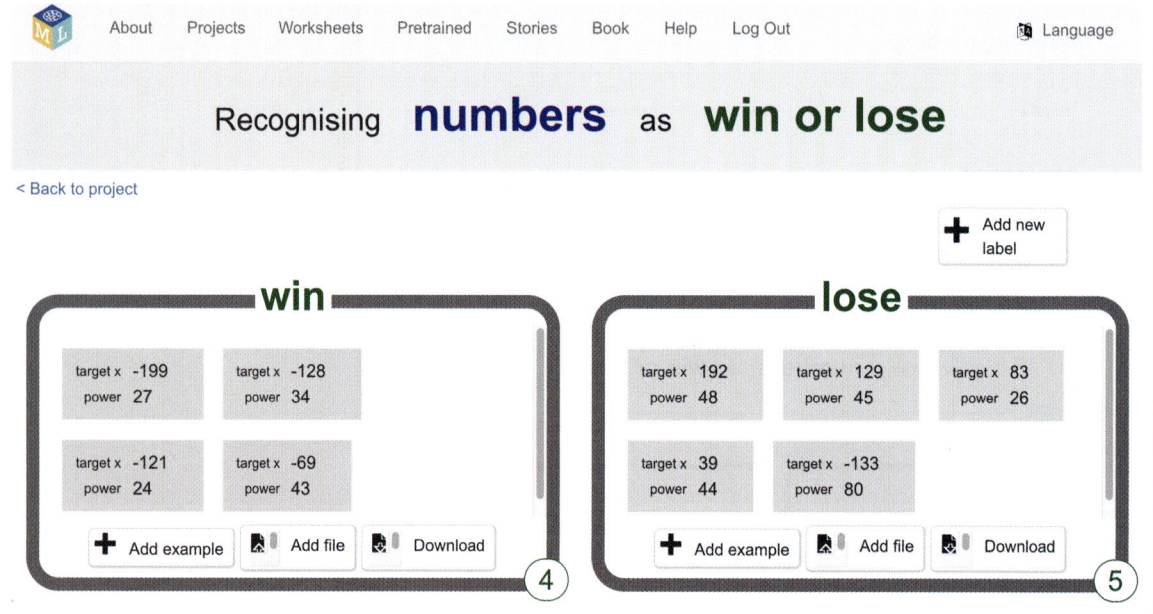

Get the machine learning model to play your game

Build 8:

1. Update the 'Launcher' sprite code to play using machine learning:

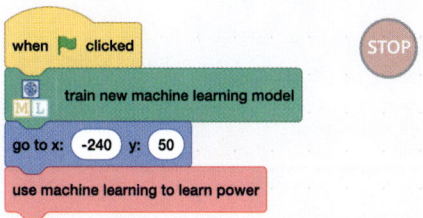

2. Update the 'use machine learning to learn power' script:

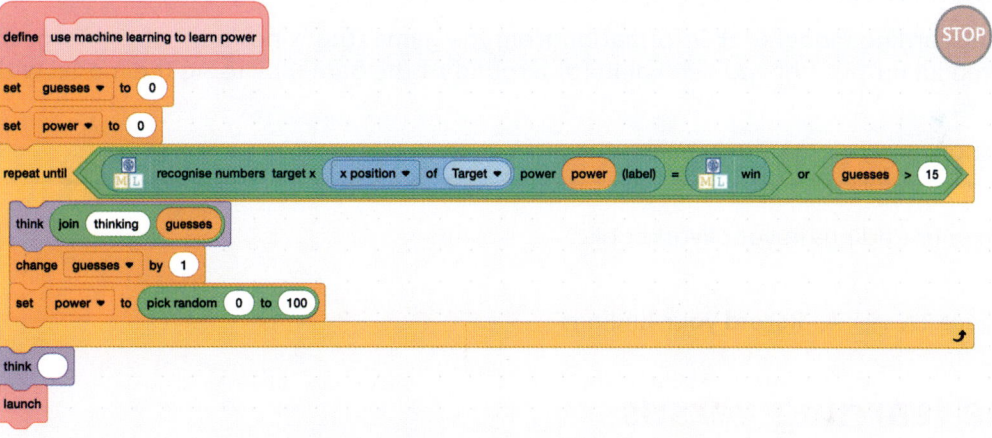

3. Click the green flag to get the machine learning model to play your game. Repeat and observe the results. Then record your results in your Workbook.

Workbook page 10: Task B, '**Record your results**'.

Reflect: Discuss: What would happen if you changed the game by moving the Target to a different position, or changed the starting position of the Projectile?

Workbook page 10: Complete the '**Reflection**' task.

Chapter 1.6 Share findings on AI and machine learning

What do we already know?
- A showcase is a time to share what you have created with an audience.
- Practising what you will say for your showcase is a good idea.

Showcase

Showcase your findings

In the last two lessons, you trained a machine learning model to play a game. The AI had to learn whether a combination of the 'x position' of the Target and the 'power' of the Launcher would result in a 'win' outcome.

The machine learning model took information from the game (the 'x position' of the Target) and produced an output (the 'power' value for a round of the game) by using the model with its training data.

In groups, showcase your results to each other.

Now write up your findings in your Workbook.

Workbook page 11: Task A, '**Record your findings**'.

Machine learning versus traditional programming

Discuss 6

How else could you have programmed a computer player for your game? Which would you prefer?

Machine learning is good for problems that have lots of data and where the patterns are difficult for humans to spot.

You also need to have good quality training data available.

The computer player for the game you made could have been programmed using traditional programming and maths. However, many problems, such as forecasting the weather, have so much data that traditional programming would take too long or be impossible.

Workbook pages 11 and 12: Task B, '**Reflection**'.

Reflect: Discuss: What kinds of problems is machine learning good for?

Well done! You have completed Chapter 1, Our digital world.

In this chapter you:

- ☑ identified ways to responsibly use technology safely, and how to protect your wellbeing
- ☑ considered how to participate responsibly in online communities
- ☑ described a range of applications of AI, including image recognition and in computer games
- ☑ trained a Scratch game to use machine learning model to play a game in Scratch.

Key terms

Artificial general intelligence – Future computer systems that might be able to perform any task in a human-like way

Community guidelines – Behaviours that are expected from members of an online community

Cookie – Small file of information that a website stores on your computer

Cyberbullying – Bullying someone using online tools

Image recognition – Identification of objects in images by computer systems using AI

Machine learning – Type of AI that uses data and algorithms to gradually produce more accurate results in a way that imitates how humans learn

Machine learning model – Algorithm plus data that can be used to make decisions or predictions

Permission – When a website or app asks to access your data or sensors

Policy – Document explaining what you can expect from a website or service, including how it will use your data

Recommendation system – Computer system that makes predictions on content that a user would like, using AI

Supervised machine learning – Type of machine learning where humans label data to help the machine learning model classify new data

Training data – Data used to train a machine learning model

Trolling – Deliberately unhelpful behaviour in an online community

Unsupervised machine learning – Type of machine learning where the AI detects patterns in data and learns how to classify new data

Voice assistant – AI system that can perform useful actions in response to voice commands

Wellbeing – The way people feel about themselves and their lives

Reflect: Discuss: How has your understanding of artificial intelligence changed during this chapter?

Workbook page 12: Complete the '**Reflection**' task.

Chapter 2 Content creation

Project: Design and record a news report that makes a prediction about future technologies

In this chapter, you will:

- create a news report collaboratively using an online word processor
- investigate the reliability of future technology predictions
- search for and use images within their usage rights
- plan, film and broadcast a future technology news report.

End of chapter project: Content creation

You will design and film something like these future technology news reports:

Chapter 2.1 Collaborative working

What do we already know?
- There are benefits in collaborating when programming.
- Digital communication makes online communities possible.
- Cybersecurity threats include data and identity theft.

Key terms
Online collaboration – People working together in apps accessed over the internet

Coworking buildings – Workspaces where people go to work together with others in the same physical location

Online collaboration tools

Online collaboration happens when people work together in apps accessed over the internet. People could be at computers in the same location, or they could be in completely different locations around the world.

There are many types of online tool that allow live collaboration, for example:

- communication platforms – video, speech or text-based spaces used for group discussion
- project management tools – apps for organising and assigning tasks, sharing files and monitoring the progress of a project
- content creation apps – word processor, spreadsheet, drawing and presentation software used for working on shared files.

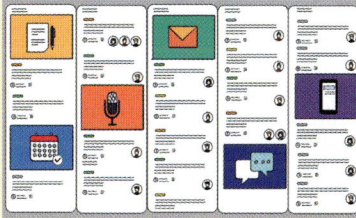
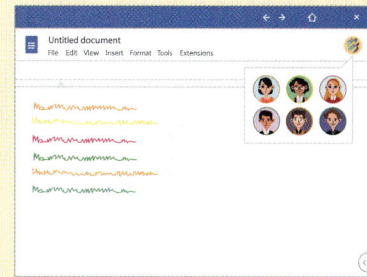

Discuss 1
When have you used online collaboration tools? Why did you use them?

Benefits of online collaboration tools include:

- Cheaper – these tools reduce the cost of **coworking buildings**, travel expenses and equipment.
- More diverse – collaboration can happen between people who can't usually work together in the same space.
- More inclusive – if everyone is using the same tools it makes it easier to train and include people.
- More efficient – they help to avoid unnecessary offline discussions, notes or email conversations in other tools.

Project: Design and record a news report that makes a prediction about future technologies

Co-authoring documents

Online collaboration tools for content creation allow automatic synchronisation. This means that everyone is able to work on the same version at the same time. This approach is often referred to as collaborative writing or co-authoring.

Online collaborative word processors have tools and features that are useful when working collaboratively.

One feature is to add comments to the document. Comments are usually added to individual words or sentences and can be seen down the right-hand side of the document. The author and time of the comment is shown alongside the actual comment. It is possible to reply to comments or to resolve them using a checkmark.

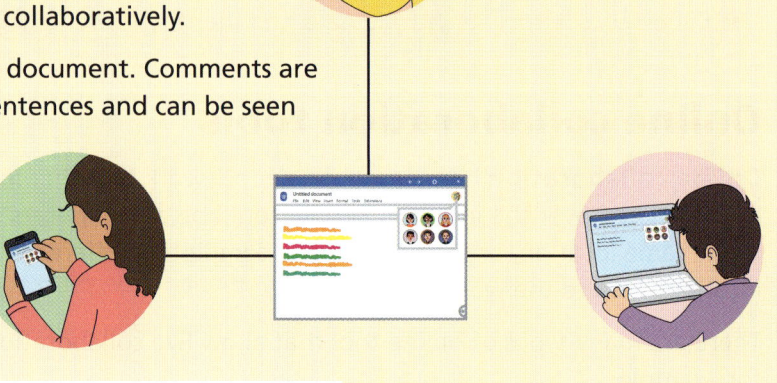

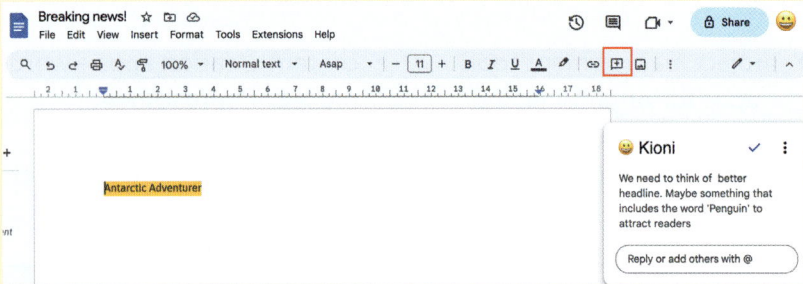

It is also possible to track changes to documents to see who last edited the document, when they edited it and what changes they made. Previous versions of a document can be viewed and restored in the document version history. It also allows you to see who else is logged onto a document at the same time you are so you can send them real-time messages and comments.

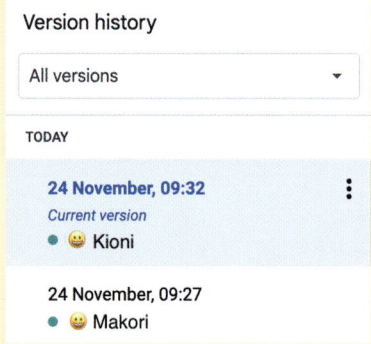

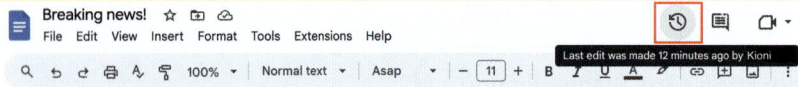

Your teacher
 Your teacher will demonstrate using an online collaborative word processor. They will show you how to add comments and track changes.

Stay safe
 One disadvantage of online collaboration tools is the risk of a cybersecurity breach, such as data or identity theft. Uninvited guests could access your files, so it is important to use strong passwords, enable any enhanced security features, and make sure you only share your files with people who need to access them.

Workbook page 13: Complete Task A, '**Breaking news!**'.

Chapter 2 Content creation

Chapter 2.1

Roles and responsibilities

When using online collaboration tools, it is important to be respectful of others. Ground rules show people how to collaborate in the best way. Ground rules for a collaborative video meeting tool could be:

- Stay on mute unless you are talking.
- Use emoji reactions to show agreement/disagreement.
- Put your hand up if you want to talk on the microphone.

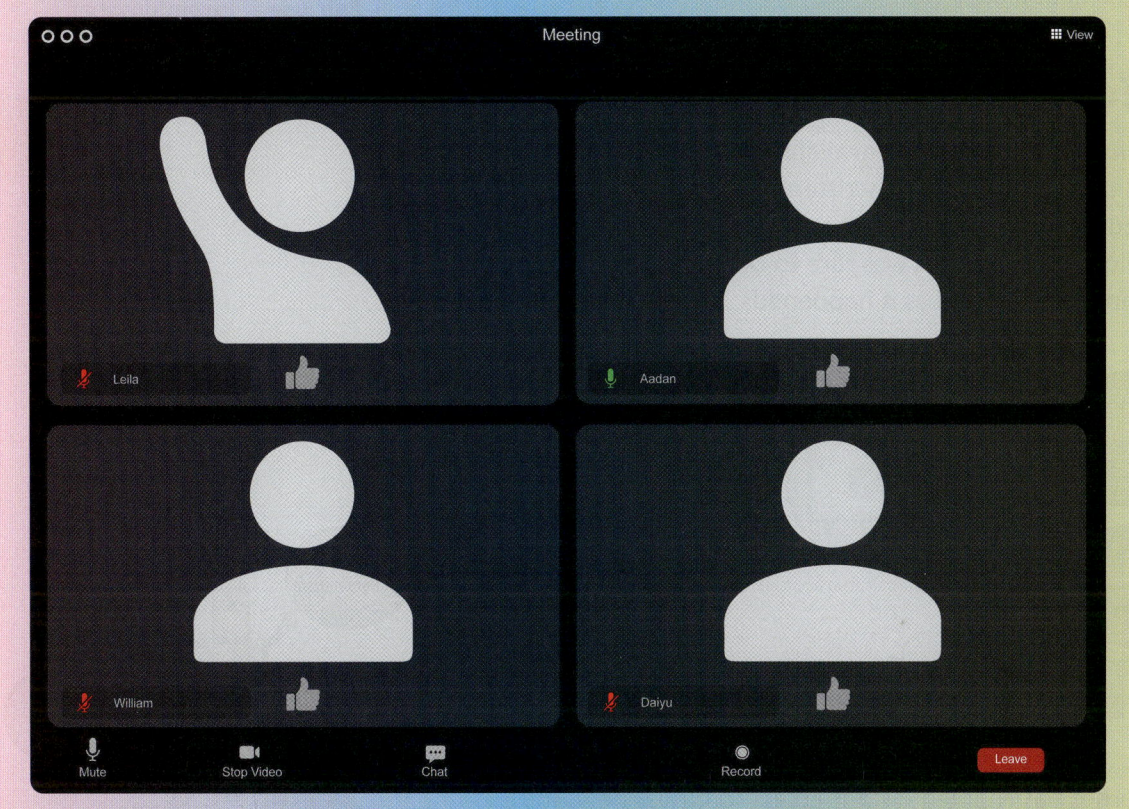

Workbook page 14: Complete Task B, '**Co-author reflection**'.

Discuss 2

What are your best ground rules for successful collaborative working in a word processor?

Reflect: If all of your collaborative work could be completed online from anywhere with a computer, why would/wouldn't you want to do it that way all of the time?

Workbook page 14: Complete the '**Reflection**' task.

Project: Design and record a news report that makes a prediction about future technologies

Chapter 2.2 Usage rights

> **What do we already know?**
> - A digital footprint is the trail of information left behind by your activity in the digital world.
> - The internet has changed the way in which people communicate, shop and access entertainment.
> - Online content can be published and accessed immediately.

> **Key terms**
>
> **Plagiarism** – Copying someone else's work then saying it is your own work
>
> **Citation** – Credit given to identify the original author of a piece of work
>
> **Fair use** – Exceptions that allow the use of work that is under copyright

Platforms for news coverage

Traditionally, news has been available via print, television and radio. News often had to be written or recorded and then delivered later in scheduled newspapers, magazines or television and radio programmes. It was rare to be able to find out what was happening in the world as it happened.

The internet has enabled instant access to global news in real time. News is now available on other non-traditional platforms. These platforms make content quick and easy to create, update and share.

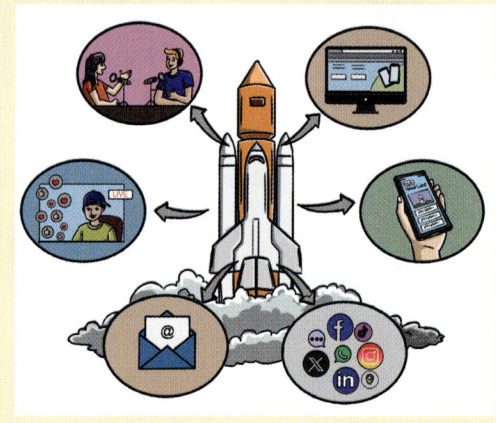

With smart devices now in many homes and vehicles, live news is available almost anytime, anywhere. People can:

- watch live news videos in social media apps on a smartphone anywhere there is a data service
- listen to live news reports on in-car entertainment systems when travelling, and personalised news summaries on a smart speaker when waking up
- receive breaking news alerts to a smartphone for installed news apps.

The availability of easy-to-access real-time coverage allows news to reach new audiences and encourages people to get more involved in the news than ever before.

Chapter 2.2

Discuss 3

Why might someone feel excluded from what is happening in the world today?

Discuss 4

How could an increase in the availability of online live news coverage have a negative effect on someone's mood?

Online live news consumption

There are positive and negative impacts of being able to access news from more sources across more platforms. Take a look at these scenarios.

The news appears in my social media feeds. I don't even have to leave the app!

If I don't know whether to trust someone online I can look at their digital footprint to see more about them.

I have more choice of where I get my news from. I avoid sources that are too controlled.

I pay an online subscription to access news. I know that the source has a good reputation for accuracy.

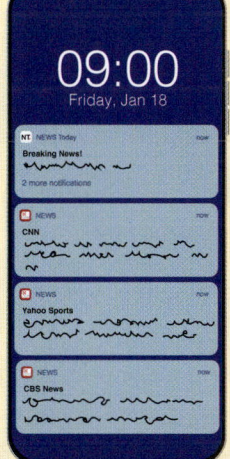

I set smartphone notifications for breaking local news, so I know instantly if and where I can help someone.

I selectively avoid the news. My local area has suffered from misrepresentation of risk.

Discuss 5

What examples have you seen where instant access to global news has been a good thing?

Project: Design and record a news report that makes a prediction about future technologies

Plagiarism, citation and fair use

Plagiarism is copying someone else's work and then saying it is a new piece of work.

Plagiarism can have legal, financial and reputational consequences and can occur in:

- music or other sounds
- images or videos
- text in books, article or speeches.

People can deliberately or accidentally plagiarise. You can check for plagiarism in your own or other people's work by:

- checking that anything inserted from somewhere else has the original author(s) properly credited
- copying part of the work into a search engine and looking for matching results
- using a plagiarism checker and making sure the plagiarism score is low.

Stay safe

! If you spot any plagiarism online, or if you feel like you might accidentally have plagiarised in work you have submitted, you should report it to your teacher.

Plagiarised work does not give credit to the original author or is used without their consent. **Citation** is a way to give credit to the original author so that people know that the information came from another source.

Citations often include the author's name, the year they published their information, the title of their work and, if possible, a link to their work.

> **Discuss 6**
> Why is citation important?

When work has a copyright it can only be used with specific permission from the original author. The laws vary from country to country and can depend on the purpose for using the piece.

Fair use exceptions allow the use of work without permission in certain situations such as criticism, news reporting, teaching or research. A court would need to decide whether or not the information used without permission is covered by fair use.

Search engine techniques

Search engine images can be filtered by licence type to find images that you are able to reuse.

> **Stay safe**
> Search engines can also be filtered using safe search settings. When searching images, this will enable you to filter out or blur explicit images.

> **Your teacher**
> Your teacher will demonstrate how to perform an image search and check the usage rights.

Workbook pages 15 and 16: Complete Task A, '**Image usage rights**'.

Reflect: Discuss: Did you find images that you wanted to use but weren't allowed to? How did you feel about this?

Workbook page 16: Complete the '**Reflection**' task.

Chapter 2.3 Future technologist reliability

> **What do we already know?**
> - You make a prediction when you think about what might happen based on your current understanding of a topic.
> - People consume news from many different online platforms.

> **Key terms**
> **Future technologist** – Expert who uses current technology trends to make predictions about the future

Future technologists

A **future technologist** (or futurologist/futurist) is an expert who researches current technology trends and developments and then uses that information to make predictions about technology in the future.

> **Discuss 7**
> Which platforms do you think a future technologist could use to reach their audience?

The role of a future technologist is to see change before it happens and then to make their predictions public before it does happen. Predictions might become reality quickly or not happen for many years; in some cases predictions will never become reality.

In the past, it was thought that:

- television would not last because people would get bored of staring at a box
- online shopping would not appeal to people because they like to get out of the house
- cars were just a fad and would be replaced once again by horses.

In 1943, the president of IBM said, "I think there is a world market for maybe five computers," and, in 1977, the chairman of the Digital Equipment Corporation said, "There is no reason anyone would want a computer in their home." Both used the information they had at the time but neither successfully predicted how computers and their use would evolve in the future.

Workbook page 17: Complete Task A, '**Researching predictions**'.

Establish trust

A lot of people like to predict the future, but that doesn't mean that everyone has done the same amount of research and investigation before making their predictions.

Discuss 8

Why is it important to look into the trustworthiness of a future technologist?

Discuss 9

How can a future technologist influence decisions now?

There are different ways to establish how trustworthy online content is:

- Check how often the author has created content.
- See whether the content or author is sponsored or has other links that might introduce bias.
- Investigate the author's credentials via 'about us' or 'contact' pages.
- Find out how many views their content has had.
- Use a search engine to search for reviews about the author.
- Make sure they have used citations for sources.
- Use a plagiarism checker or paste an extract into a search engine.

Workbook page 18: Complete Task B, '**Establish trust**'.

Discuss 10

How trustworthy were the predictions you found?

Reflect: Share the prediction that is most likely to happen based on the research you have carried out.

Workbook page 19: Complete the '**Reflection**' task.

Project: Design and record a news report that makes a prediction about future technologies

Chapter 2.4 — Plan a news report

What do we already know?

- Online word processors allow people to work together via the internet.
- Videos can be used to present news reports.
- Interviews are a tool used to get information from experts.

Key terms

News script – Text that details dialogue, timing and camera actions of a news item

Camera direction – Shots and angles used to describe the position, view and movement of a camera

Project brief

Design and record a news report that makes a prediction about future technologies

You will use video to record a news report. The news report will be scripted so that everyone knows their part before filming.

The news report will be an interview with:

- a presenter to introduce the interview
- an interviewer to ask questions
- an interviewee expert to answer questions
- a camera person to record the news report.

Your teacher

 Your teacher will show you an interview-style news report.
You will use an online collaborative word processor to create the script.

Discuss 11

Why would collaborative tools be useful for creating a script?

Discuss 12

What ground rules aid collaborative working?

Workbook page 20: Complete Task A, '**News report overview**'.

Writing a script

When writing a **news script**, you should clearly label who will be speaking each part. You should also include any **camera directions** and use of props.

> **Tip**
> Often, experts interviewed will not need fully scripted responses as they know the topic extremely well. As you are not experts in the topic of your news report, you should script the interview answers in full, or as bullet points, to help when you record the report.

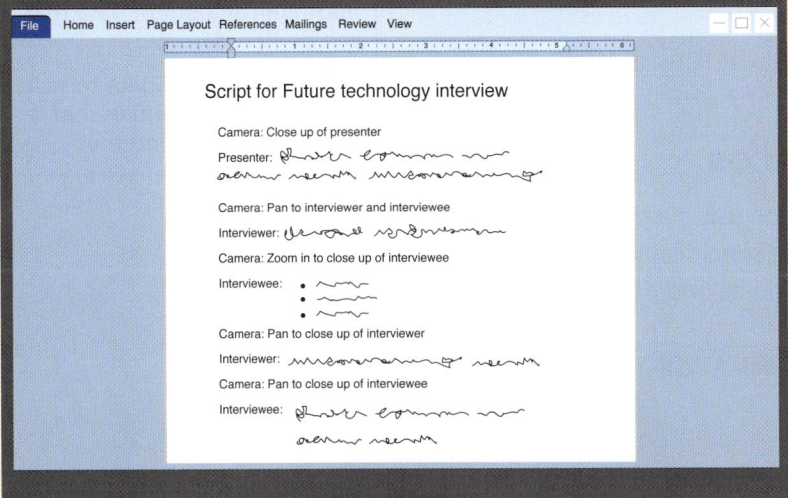

> **Tip**
> When writing your script, you should frequently read it aloud. You will find that a script can sound great in your head as you read it, but when you say it aloud you will want to make some small changes.

Workbook pages 20 and 21: Complete Task B, '**Interview observations**'.

Build 1:
As a group, create a document in your online word processor and share it between you so that you can work collaboratively on your script. Use the notes in your Workbook to help to write the script.

Workbook pages 21 and 22: Complete Task C, '**News report notes**'.

Reflect: Discuss: What were the advantages and disadvantages of working collaboratively on the script?

Workbook page 22: Complete the '**Reflection**' task.

Project: Design and record a news report that makes a prediction about future technologies

Chapter 2.5 Record a news report

> **What do we already know?**
> - A storage hierarchy can be used to save and find files.
> - Computer systems use a range of storage devices.
> - Different types of files have different sizes.

> **Key terms**
> **Storage capacity** – The maximum amount of data that can be stored
> **Cloud storage** – Space to save, access and manage files that is accessed over the internet

Video recording devices

To create your news report, you will be using a device capable of recording videos. Some devices are specifically built to record videos but others have video recording as one of many functions.

Devices have limits to their **storage capacity**. Video files can be very large and so take up a lot of storage space.

Different video recording devices store video files in different ways. Some devices use:

- external storage cards that can be inserted into the video recording device; these can be removed and inserted into other devices to transfer files
- a data cable connected to another device; the files can be stored on the video editing device before transfer or sent directly to be stored on the other device
- an internet connection to store files in **cloud storage** that can be accessed by other devices connected to the internet.

> **Your teacher**
> Your teacher has used your device to record some short videos and will now show you the file sizes.

Recording your project

You will record your news report as one continuous unedited video and then transfer your best recording to the school network.

> **Your teacher**
> Your teacher will now demonstrate how to use your device to record, transfer and delete videos.

Workbook page 23: Complete Task A, '**Using the video recorder**'.

Build 2:
Record your video.

Workbook page 24: Complete Task B, '**Video notes**'.

Workbook page 25: Complete Task C, '**My video**'.

Reflect: Why might it have been better to record the video in more than one take and then edit it together later?

Workbook page 25: Complete the '**Reflection**' task.

Tip

Tips for recording your video:
- Rehearse your video before filming.
- Remember to use your props, but make sure they can be clearly seen on camera.
- Stick to the script during recording to respect others in your group.
- Set your scene. If possible, make sure you are filming in an area with a nice background and good light.
- Consider the sound recording quality and what could impact that; for example, distance from the microphone, wind or background noise.
- Think about your tone, speed and body language.
- Play your recording back and watch it. Make sure everyone is happy with it.

Project: Design and record a news report that makes a prediction about future technologies

Chapter 2.6　Broadcast a news report

What do we already know?

- The role of a future technologist is to be an expert who uses current technology trends to make predictions about the future.
- Plagiarism is taking someone else's work and saying that it is your own.
- News can be reported through video interviews.

Movie time

This lesson will take the form of a future technologies news program.

Showcase

Everyone's news reports will be broadcast to the class.

Discuss 13

After each video, discuss: How realistic is the future technology prediction?

Workbook page 26: Complete Task A, '**Reflection**'.

Congratulations!

Well done! You have completed Chapter 2, Content creation.

In this chapter you:

- ☑ created a news report collaboratively using an online word processor
- ☑ investigated the reliability of future technology predictions
- ☑ searched for and used images within their usage rights
- ☑ planned, filmed and broadcast a future technology news report.

Key terms

Camera direction – Shots and angles used to describe the position, view and movement of a camera

Citation – Credit given to identify the original author of a piece of work

Cloud storage – Space to save, access and manage files that is accessed over the internet

Coworking buildings – Workspaces where people go to work together with others in the same physical location

Fair use – Exceptions that allow the use of work that is under copyright

Future technologist – Expert who uses current technology trends to make predictions about the future

News script – Text that details dialogue, timing and camera actions of a news item

Online collaboration – People working together in apps accessed over the internet

Plagiarism – Copying someone else's work then saying it is a new piece of work

Storage capacity – The maximum amount of data that can be stored

Reflect: Discuss: Share something you really liked about a news report video made by someone else.

Workbook page 27: Complete the '**Reflection**' task.

Chapter 3 — Create with code 1

Project: Design a feedback kiosk using a flowchart and build it in Scratch

 In this chapter, you will:

- describe the role of logic gates in circuits
- use logic statements
- predict the outcome of flowcharts
- debug flowcharts
- design, build and showcase a flowchart and Scratch project for a feedback kiosk.

End of chapter project: Design a feedback kiosk using a flowchart and build it in Scratch

You will design something like these projects:

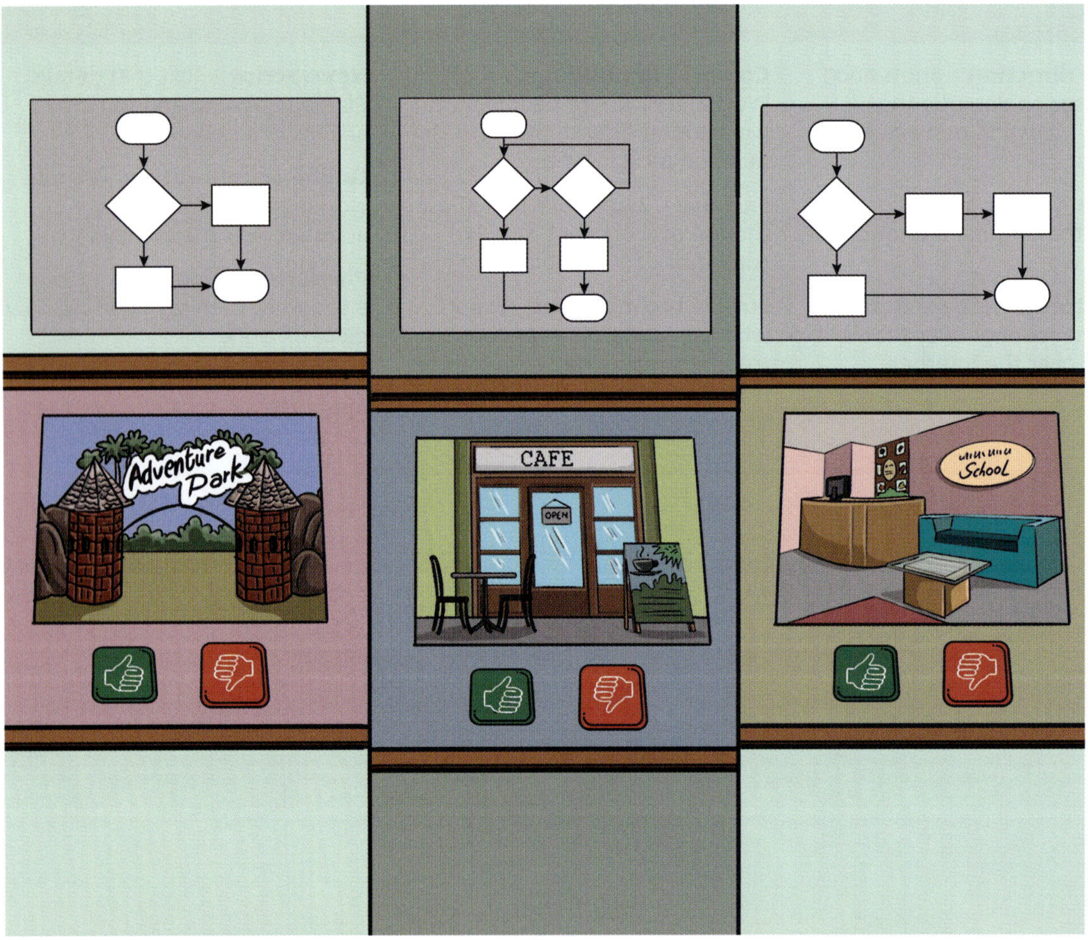

Chapter 3.1 Logic gates

> **What do we already know?**
>
> - Computers follow explicit instructions that are based upon logic.
> - Computers represent data in binary (0s and 1s).
> - The symbols used in flowcharts.
> - How to follow an algorithm presented as a flowchart.

> **Key terms**
>
> **Boolean logic** – Form of algebra that uses the operators AND, OR and NOT
>
> **Boolean expression** – Expression that either has a TRUE or FALSE outcome
>
> **Logic gates** – Used in logic circuits in computers. They are based upon the Boolean logic operators AND, OR and NOT

Boolean logic

Boolean logic uses the operators AND, OR and NOT. In computer science, you can write a **Boolean expression** that uses these logic operators. Boolean expressions always result in either TRUE or FALSE.

Here are some example Boolean expressions that use the operators AND, OR and NOT.

AND

'The temperature is <u>over 20</u> degrees AND it is <u>raining</u>'

For this statement to be TRUE, *both* conditions need to be TRUE.

> **Discuss 1**
>
> Why is the image on the right FALSE?

> **Discuss 2**
>
> There are two other possible scenarios for this Boolean expression. What are they?

OR

When the OR operator is used, the whole statement is TRUE when *one* or *both* statements are TRUE.

For example: 'The snowman has a hat OR The snowman has a scarf'

> **Discuss 3**
>
> There are two other scenarios for this Boolean expression. What are they?

Project: Design a feedback kiosk using a flowchart and build it in Scratch

NOT

The NOT operator takes a single condition (input). If that condition is FALSE it returns TRUE; if that condition is TRUE it returns FALSE.

For example: NOT 'It is raining'

Your teacher

 Your teacher will now give you an activity where you will respond to some Boolean expressions.

Workbook page 28: Complete Task A, '**Boolean expressions**'.

Logic gates

Computers represent data using 1s and 0s. The values 1 and 0 are used to represent the flow of an electrical current through a circuit. 1 represents ON, which means that the current can flow, and 0 represents OFF, which means that the current cannot flow. The numbers 1 and 0 also represent the values TRUE and FALSE.

| 1 | ON | TRUE |
| 0 | OFF | FALSE |

Boolean logic is applied to computer circuits through **logic gates**. A logic gate takes an input (electrical current) and either stops the current or allows it to flow based on the input and the type of logic gate.

The AND logic gate

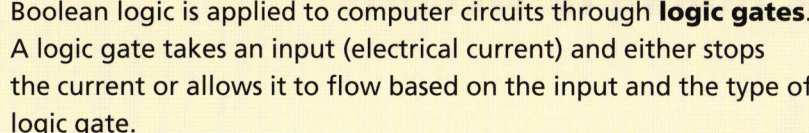

The AND logic gate takes two inputs, just like our example earlier. An AND gate will allow the current to flow if *both* inputs are ON or TRUE.

The OR logic gate

The OR logic gate takes two inputs. An OR gate will allow the current to flow if *one* or *both* inputs are ON or TRUE.

The NOT logic gate

The NOT logic gate takes one input. If the input is ON/TRUE then the output will be OFF/FALSE. If the input is OFF/FALSE then the output will be ON/TRUE.

> **Workbook** pages 28 and 29: Complete Task B, '**Investigate logic gates**'.

> **Reflect:** Think of your own real-world example of a Boolean expression and share it with a partner. Check that your Boolean expression can only result in a TRUE or FALSE response.

> **Workbook** page 29: Complete the '**Reflection**' task.

Project: Design a feedback kiosk using a flowchart and build it in Scratch

Chapter 3.2 Selection in flowcharts

> **What do we already know?**
> - The Boolean logic operators AND, OR and NOT can be used to write Boolean expressions that result in a TRUE or FALSE outcome.
> - Flowcharts can be used to express an algorithm.
> - Flowcharts use specific symbols to match the type of process that is happening at each stage.

Boolean operators in Scratch

Boolean operators can be used in programming as well as logic circuits. Here is an example of a Boolean AND operator being used in a Scratch program:

This group of blocks states that 'If the time variable is greater than 10 AND the score variable is greater than 5 then call (run) the jump subroutine'.

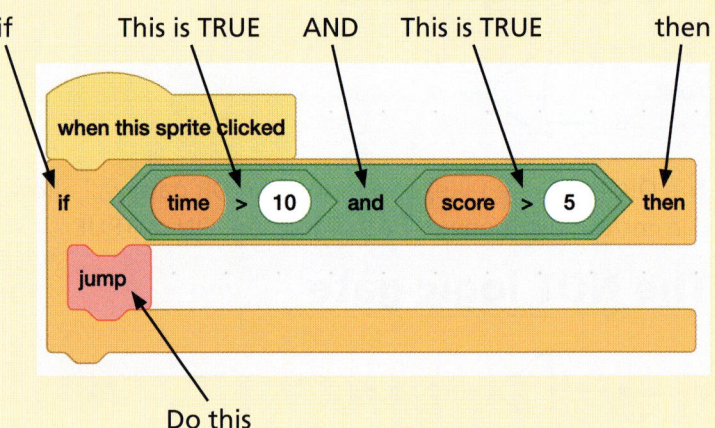

Workbook page 30: Complete Task A, '**Label the flowchart symbols**'.

Boolean operators in flowcharts

You can also use Boolean operators in a decision (diamond symbol) in a flowchart. The flowchart below uses the AND operator.

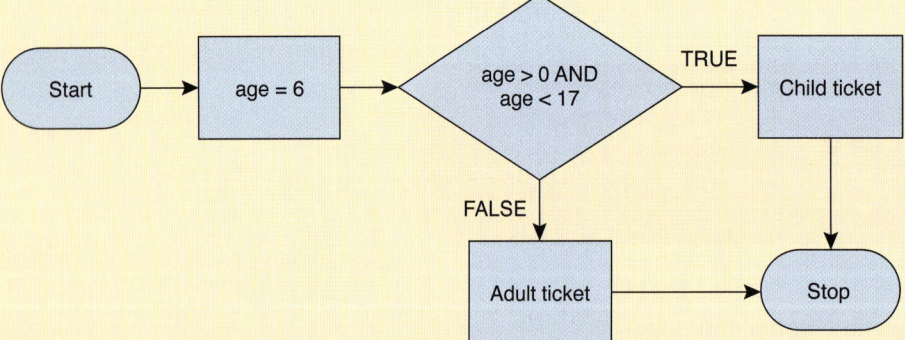

> **Discuss 4**
> What would be the OUTPUT of this flowchart?

The correct ingredients

A student is designing a Scratch game that requires a player to enter ingredient combinations to make some healthy food. If the player uses the correct types of ingredients, the bowl will contain a pleasant soup. If the player has the wrong types of ingredients, the bowl will contain stinky slime.

The student has created a flowchart to plan their code.

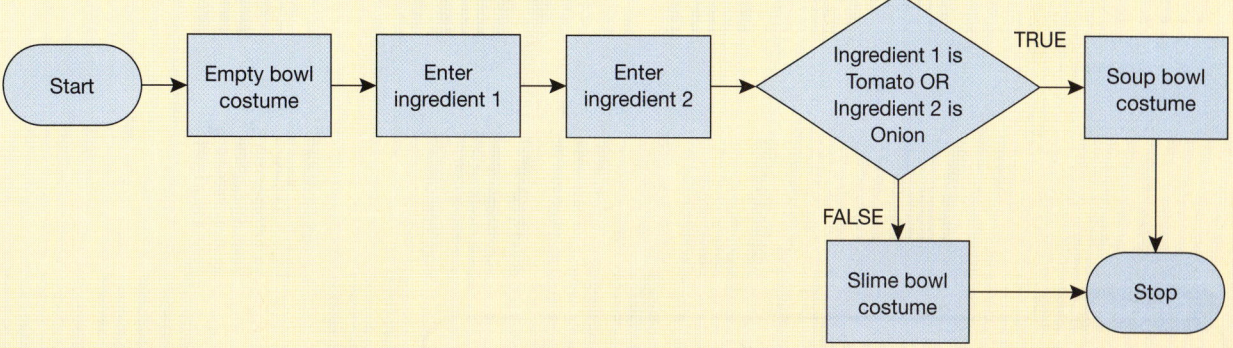

Discuss 5

What will happen if the first ingredient is a tomato and the second ingredient is a potato?

Workbook pages 30 and 31: Complete Task B, '**The correct ingredients**'.

Your teacher

Your teacher will show you the Scratch project that the student made using the flowchart. Check whether your predictions were correct by running the project and putting a tick or cross next to each answer in the table on page 31 of your Workbook.

Reflect: Discuss: Were any of your predictions incorrect? If so, which ones and what was the actual output?

Workbook page 31: Complete the '**Reflection**' task.

Chapter 3.3 Debug flowcharts

What do we already know?

- To debug code, you find and fix any errors within it.
- Flowcharts are used to design the logic for a program.

Identify the error

A student has designed a flowchart for a Scratch program that will help them to decide whether or not they need to wear a raincoat. They have used the flowchart below to create the Scratch program but it isn't working as expected. It is meant to say that they need to wear a raincoat when it is raining, but it doesn't.

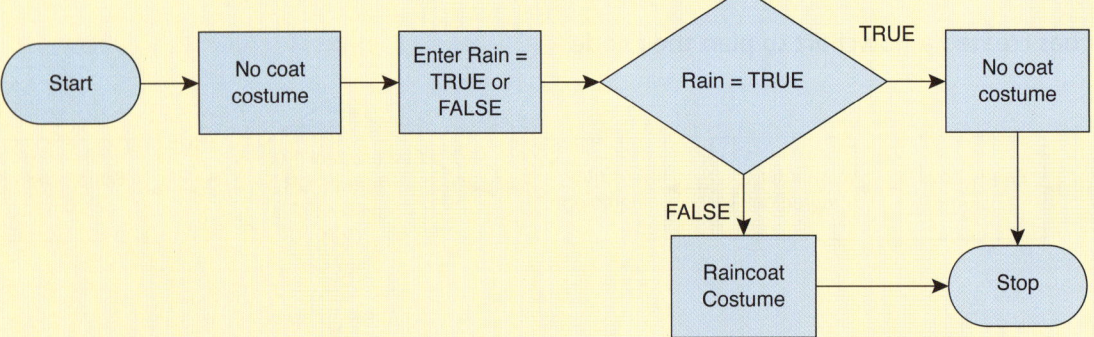

Discuss 6

Where is the error? How should they fix the error?

Workbook pages 32 and 33: Complete Task A, **'Debug the flowcharts'**.

Reflect: These flowcharts will only check for a condition once and then the program will stop. What could be introduced to the flowchart to make the program constantly check for changes in the condition?

Workbook page 33: Complete the **'Reflection'** task.

Chapter 3.4 Design a flowchart

What do we already know?

- Flowcharts can be used to design programs.
- Flowcharts use different symbols to show different types of processes that the code could action.
- Flowcharts can be converted into program code.

Project brief

For your project, you will design a feedback kiosk using a flowchart and then build it in Scratch, using your flowchart as a guide. Your feedback kiosk could be for:

- parents and carers in the school reception area
- learners in the classroom to rate how well they understood the lesson
- your favourite cafe, shop or activity space
- anywhere you visit regularly.

The kiosk should ask the user to provide a rating by entering a number out of 10. A sprite should then react based on the number entered. You can choose how you want your sprite to react. It could be through switching costumes, costume effects or performing an animation.

Your program should:

- ask for a user rating between 1 and 10
- provide a sprite reaction for a low number
- provide a sprite reaction for a neutral number
- provide a sprite reaction for a high number.

Here is an example algorithm for rating a playground that uses costume changes for reactions:

```
Ask for user rating
If rating < 4 then
    Sad face
If rating > 3 and rating < 7 then
    Neutral face
If rating > 6
    Happy face
```

This algorithm can then be presented as a flowchart in the following way:

Project: Design a feedback kiosk using a flowchart and build it in Scratch

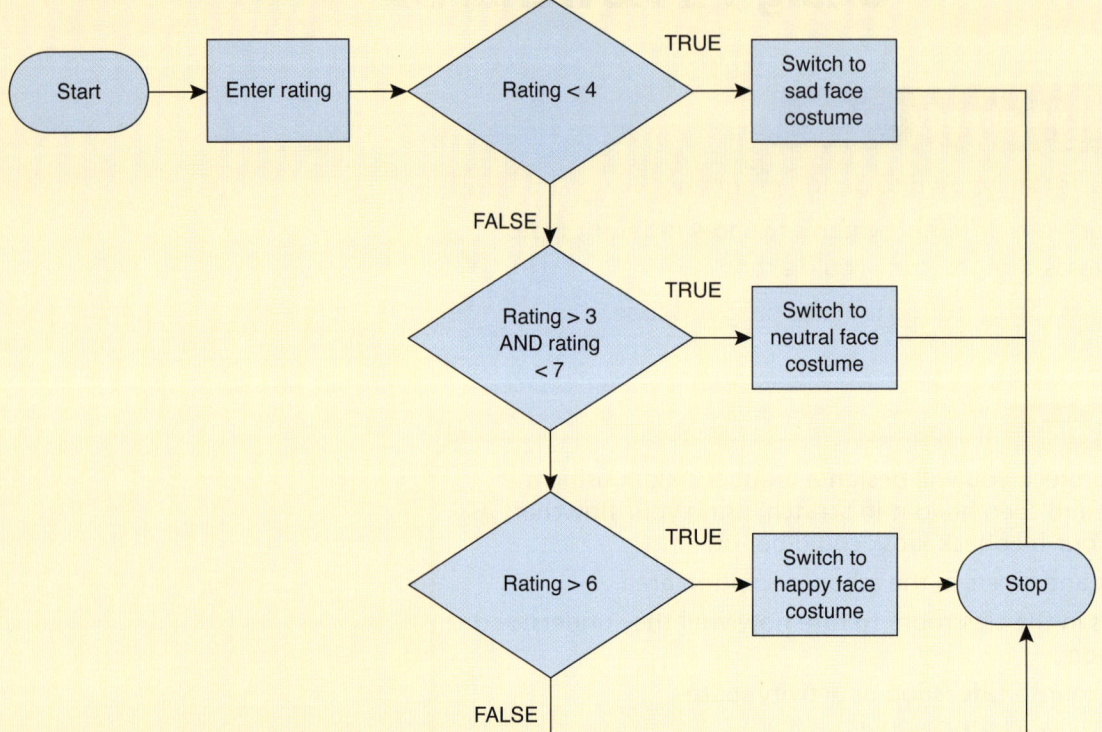

The flowchart can then be implemented in Scratch to produce a working project.

The Scratch blocks are shown on the right.

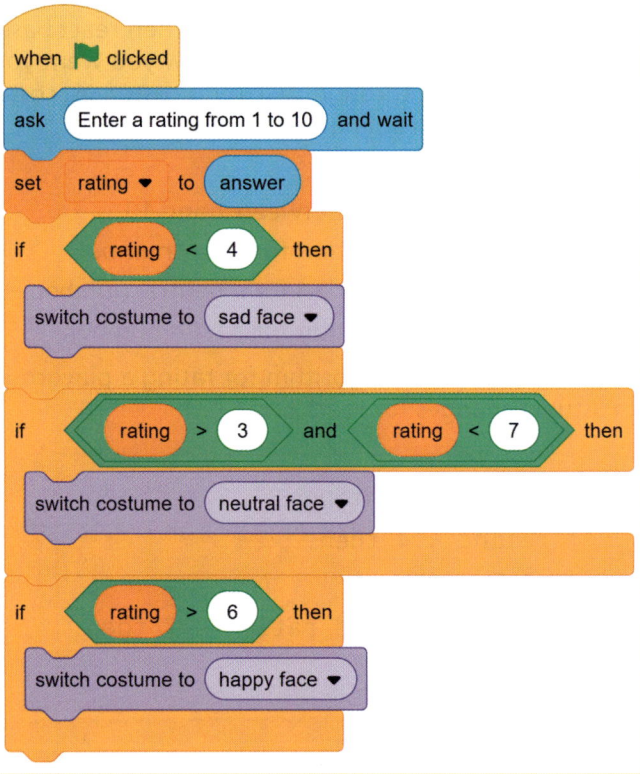

> **Your teacher**
>
> ! Your teacher will give you some time to explore the example project that uses this code.

Workbook page 35: Complete Task A, '**Your feedback kiosk**'.

Reflect: Share your design ideas with a partner and ask for feedback on your idea.

Workbook page 35: Complete the '**Reflection**' task.

Chapter 3.5 — Build a program from a flowchart

What do we already know?

- Flowcharts can be converted into program code.

Build 1:

Build your Scratch program using your design.

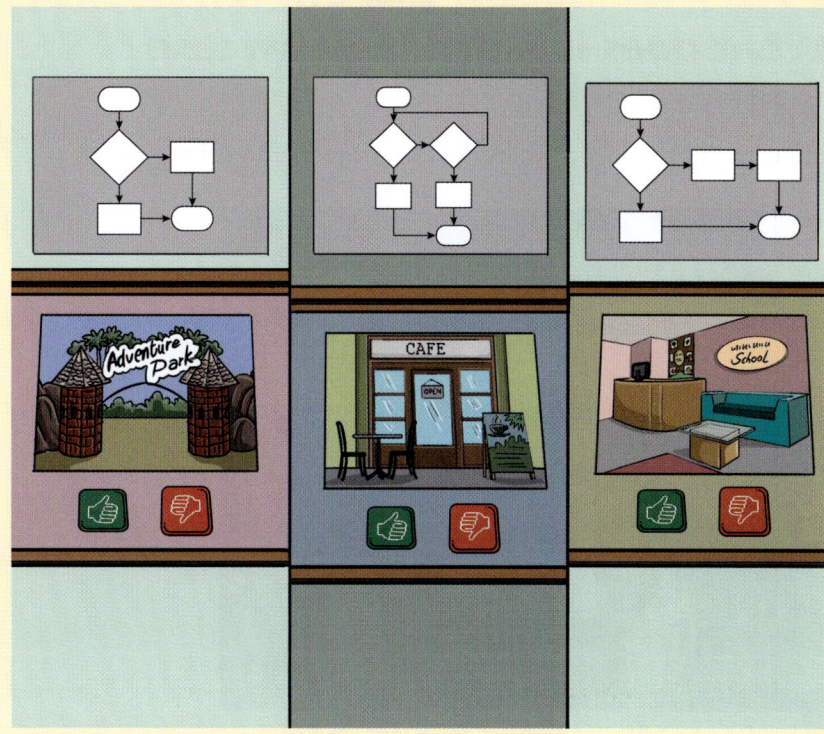

In the last lesson, you designed your feedback kiosk in your Workbook. Now you will use your design to build your Scratch program.

Remember that your program should:

- ask for a user rating between 1 and 10
- provide a sprite reaction for a low number
- provide a sprite reaction for a neutral number
- provide a sprite reaction for a high number.

Your sprite can react in many different creative ways. It could be:

- a costume change
- a costume effect
- an animation
- a sound
- something else.

Workbook page 36: Complete Task A, '**Test your program**'.

Reflect: Did your program work as intended? If not, did you fix any errors?

Workbook page 36: Complete the '**Reflection**' task.

Project: Design a feedback kiosk using a flowchart and build it in Scratch

Chapter 3.6 Showcase a flowchart and a program

> **What do we already know?**
> - A showcase is a time to share what you have created with an audience.
> - Practising what you will say for your showcase is a good idea.

What does it mean to showcase your project?

Showcasing is a way of sharing your project with other people.

Tips for showcasing your project:

- Speak loudly and clearly.
- Open your Workbook at the page with your flowchart on so that it is ready to share.
- Allow your user to click the green flag to see what happens and enter a chosen number.

> **Showcase**
>
> Showcase your feedback kiosk. You should:
> - State the location for your kiosk.
> - Say why you chose the reactions that you did.
> - Give your audience time to look at your flowchart and compare it to the code in your Scratch project.

Workbook page 37: Complete Task A, '**Reflection**'.

Well done! You have completed Chapter 3, Create with code 1.

In this chapter you:

- ☑ described the role of logic gates in circuits
- ☑ used logic statements
- ☑ predicted the outcome of flowcharts
- ☑ debugged flowcharts
- ☑ designed, built and showcased a flowchart and Scratch project for a feedback kiosk.

Key terms

Boolean expression – Expression that either has a TRUE or FALSE outcome

Boolean logic – Form of algebra that uses the operators AND, OR and NOT

Logic gates – Used in logic circuits in computers. They are based upon the Boolean logic operators AND, OR and NOT

Reflect: Share your favourite project made by someone in your class.

Workbook page 37: Complete the '**Reflection**' task.

Chapter 4: How computers work

Project: Create a collaborative pixel art installation that represents your local community

In this chapter, you will:

- explore automation in the health and advertising industries
- compare applications software with systems software
- discover how binary is used to represent numbers, images, characters and sound
- create a collaborative piece of pixel art that represents your local community.

End of chapter project: Pixel art

Create a collaborative pixel art installation that represents your local community

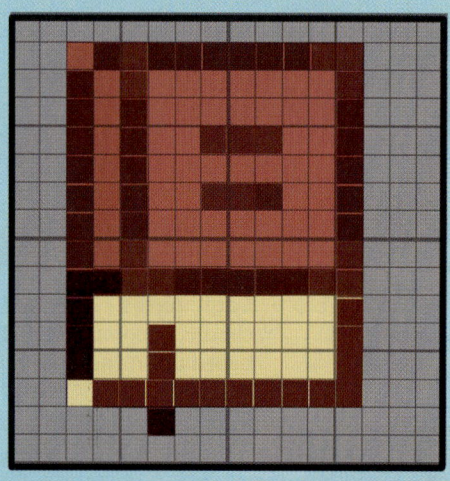

Chapter 4.1   Software and automation

What do we already know?

- Software is used to provide instructions for computer hardware. Examples of software are Microsoft Word and Google Chrome.
- Software needs specific hardware to work correctly. For example, a game with high-quality graphics requires that the computer has a high-quality graphics card.

Key terms

Application software – Software used for a specific task, such as a word processor

System software – Software that controls how the computer operates

Operating system – Software that manages the hardware, software and resources of a computer

Utility software – Software that provides additional support for a system, such as antivirus or disk cleanup management

Automation – Carrying out tasks with little to no human interaction

Application software

Application software is a type of software that is used to complete a specific task. For example, if you would like to write a book or a letter then you would use word processing software, which is a type of application.

Applications are often referred to as 'apps' and can be installed on a computing device as needed. Other types of application software are:

- Spreadsheets (Excel®, Google Sheets™)
- Web browsers (Chrome™, Safari®)
- Video editors (Adobe Premiere®, Clipchamp®)
- Sound editors (Audacity®, Sound Forge®)
- Desktop publishers (Canva™, InDesign®)

Discuss 1
Can you think of any more examples of application software?

Project: Create a collaborative pixel art installation that represents your local community

System software

System software works directly with the hardware of the computer and controls how the computer operates. An application needs system software to work.

An example of system software is the **operating system**. The operating system manages the hardware, software and resources of a computer. Operating systems that you might be familiar with are:

- Microsoft Windows
- Linux
- MacOS

Another type of system software is **utility software**. Utility software provides additional support for a system, such as antivirus or disk cleanup management.

Workbook page 38: Complete Task A, '**System or application software**'.

Automation

Software can be used to provide **automation** tools to make processes more efficient. Automation tools carry out a task on behalf of a user with little or no supervision.

Chapter 4.1

Automation in the health industry

The health industry uses automation tools to make their processes more efficient. Here are some examples of where automation might be used in the health industry:

- sending a text message reminder about an upcoming appointment
- medication warnings for a doctor when prescribing medication
- automatically scheduling appointments based on urgency and time waiting
- automatically adjusting staffing levels to meet times of high demand for doctors and nurses.

Discuss 2

Do you think that it is a good idea to leave automation software to make decisions about the urgency of a patient's illness?

Project: Create a collaborative pixel art installation that represents your local community

Automation in the advertising industry

The advertising industry's main objective is to target potential customers with adverts. Automation can help with this by tracking a user's searches, location and purchases online. Tools can use this data to target specific adverts to customers who are likely to buy a product.

Here are some other examples of automation in the advertising industry:

- scheduling social media posts for days and times where users are most likely to read them
- targeting adverts to specific demographics, for example women aged 25–30
- testing multiple thumbnails for an online video to see which one gets the most traffic (views).

Discuss 3

Have you ever been on a webpage and seen an advert that is linked to what you have recently searched for? Do you think this is good or bad?

Stay safe

 Automation tools use cookies to track what you are doing online. Accept only essential cookies if you don't want your data to be used in this way.

Workbook pages 38 and 39: Complete Task B, '**Is automation good or bad?**'.

Your teacher

 Your teacher will now ask you to debate automation with the class.

Reflect: Which side did you agree with the most? Can automation only be good or bad? Could it be both?

Workbook page 39: Complete the '**Reflection**' task.

Chapter 4.2 Binary representation

What do we already know?
- Computers represent data in binary (0s and 1s).

Binary representation of numbers

Humans use a number system called **decimal** or **denary**. The decimal system is based on ten digits, 0–9. Historians believe this is because humans have ten fingers on their hands and these were used for counting.

The decimal system is referred to as a 'base-10' number system because it has 10 digits.

When you count in the decimal system you use up all the digits in order. When you reach 9 there are no digits left to use so you start again.

You can see this in the table below. When you count past 9, the right column resets to 0 and an additional digit is used which is now worth one 10.

10	1
	8
	9
1	0

Key terms

Decimal – Base-10 number system that uses 10 digits, from 0 to 9

Denary – Alternative word for 'decimal'

Characters – Numbers, letters and symbols, for example, A, " or 9

Analogue sound – Sounds that are all around us in the non-digital world

Computers use switches to represent data and these are either on (1) or off (0).

When you count past 1 in binary, the left-hand digits begin again from 0 and an additional digit is used which is now worth one 2.

2	1
	0
	1
1	0

Tip
Notice that column 1 is headed 2 and not 10 as in the decimal system.

Your teacher
 Your teacher will now demonstrate counting from 1 to 10 in binary using a Scratch project.

Workbook page 40: Complete Task A, '**Count in binary**'.

Binary representation of characters

Your keyboard allows you to type a range of **characters** as input into your computer. They are then displayed on your screen as output. Each tap on your keyboard sends an electrical signal to your computer that represents a series of 0s and 1s.

For example, a standard English or American keyboard has a capital F on it. The binary code for F is 01000110. The 0s and 1s (electrical signals) are processed by your computer and displayed on the screen as the letter F.

Binary representation of sound

Analogue sound can be represented by a sound wave like the one in this image.

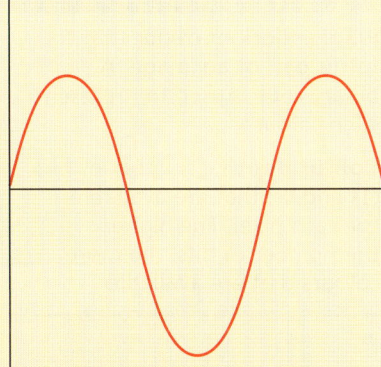

When sound is processed by a computer it is converted to a digitised sound. Digitising a sound involves assigning a binary code to sections of the sound wave.

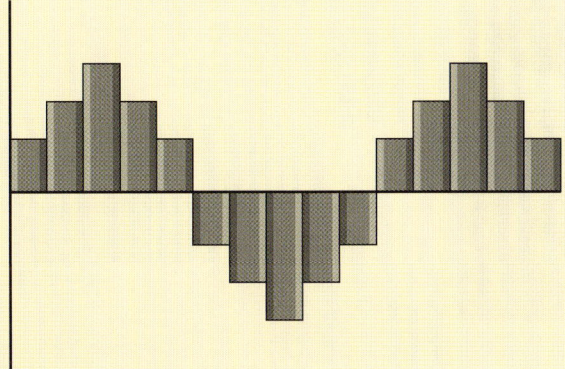

Samples of the wave are taken and a binary number is used to represent that sample. The more frequently the sound is sampled, the better the quality of the sound. However, a higher quality sound leads to an increase in file size due to the increased number of samples.

> **Workbook** page 41: Complete Task B, '**Represent characters in binary**', and pages 41 and 42: Task C, '**Represent sound in binary**'.

> **Reflect:** In the next lesson you will discover how images are represented using binary. Discuss how a binary number could be used to represent a single colour.

> **Workbook** page 42: Complete the '**Reflection**' task.

Chapter 4.3   Represent images in binary

What do we already know?

- Computers represent data in binary (0s and 1s).

Pixels in an image

If you magnify a digital image, you can see that it is made up of tiny blocks of colour. These blocks of colour are known as **pixels**.

Key terms

Pixel – Block of colour that forms part of an image. A binary code is used to represent each colour

Colour depth – Number of bits (1s and 0s) that are available for each pixel. The higher the number of bits, the more colours that are available

Pixels and binary codes

Each pixel can be represented by a binary code. Here is an example of an image that uses just one binary digit to represent a colour. A '1' is used to represent blue and a '0' is used to represent white.

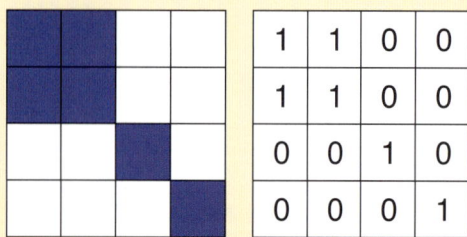

Each pixel is represented by 1 binary digit. This means that it has a **colour depth** of 1 bit (binary digit).

Workbook page 43: Complete Task A, '**Draw the 1-bit colour depth images**'.

More colours, more bits!

Here is an example of a 2 bit colour depth image. By having 2 bits, there are now more options for colours.

There are four combinations of 1s and 0s that can be represented with two bits.

| 00 |
| 01 |
| 10 |
| 11 |

This means that you can now create an image that uses four colours.

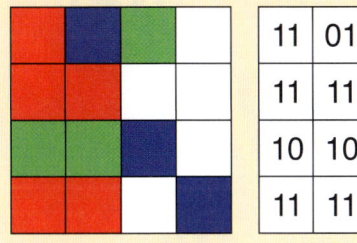

With more bits, or a higher colour depth, you can create better-quality images that use more colours.

Workbook page 44: Complete Task B, '**Colour combinations**'.

Project brief

Create a collaborative pixel art installation that represents your local community

Your project for this chapter is to create your own piece of pixel art that will be joined together with all of the art pieces made in your class to create one giant piece of pixel art. The pixel art will use a maximum of eight colours. This means that it will have a colour depth of 3 bits.

Your pixel art image must represent something in your local community. It could be:

- a school badge image
- a school mascot
- a famous building, bridge or statue near your school
- anything that represents your community. Maybe your area is famous for inventing something unique – that would be a good thing to celebrate.

Discuss 4

What images represent your local community?

Workbook page 45: Complete Task C, '**Images that represent my community**'.

Reflect: What image do you think you might use in your project?

Workbook page 45: Complete the '**Reflection**' task.

Chapter 4.4 Design your pixel art

What do we already know?

- Images are represented in binary by assigning a bit pattern to each pixel.
- The number of colours that can be used in an image depends on the colour depth of the image. The higher the colour depth, the more colours can be used.

Your pixel art

In the last lesson, you discussed different images that represent your school community and added some sketches and ideas to your Workbook.

Your pixel art will have a colour depth of 3. This means that your image can use a maximum of 8 colours. It is important that you choose these colours carefully to help you to create the best image that you can.

Here are some potential colour palette options that you could use for your pixel art:

Palette 1 (ice)

000	001	010	011	100	101	110	111
Black	Light grey	Light blue	Blue	Dark blue	Light purple	Purple	White

Palette 2 (sunset)

000	001	010	011	100	101	110	111
Black	Dark purple	Purple	Pink	Orange	Yellow	Light yellow	White

Palette 3 (pastel)

000	001	010	011	100	101	110	111
Black	Light pink	Light orange	Yellow	Light green	Light blue	Light purple	White

These are just options. You can design your own colour palette if you would like to. Just make sure that you have access to the colours that you need.

Workbook pages 46 and 47: Complete Task A, '**Design three versions of your pixel art**'.

Reflect: Do you have a favourite design from the three that you drew in this lesson? Were you happy with your colour palette? Did you have to change your colour palette as you were drawing the image?

Workbook page 47: Complete the '**Reflection**' task.

Project: Create a collaborative pixel art installation that represents your local community

Chapter 4.5 Create your pixel art

Your final design

In the last lesson, you designed three versions of your pixel art. You also spent some time reflecting on your designs. This lesson, you should decide which design you will use for your final art piece. Make sure that your chosen design:

- has a suitable colour palette
- has the correct binary codes written in each square
- represents your local community.

> **Your teacher**
> Your teacher will tell you where to draw your final design. This could be on page 48 of your Workbook or on a larger piece of paper.

Workbook page 48: Complete Task A, '**Draw your final design**'.

Build 1:
Create your final design.

Reflect: How does your artwork represent your school community?

Workbook page 48: Complete the '**Reflection**' task.

Chapter 4.6 Showcase your artwork

What do we already know?

- A showcase is a time to share what you have created with an audience.
- Is a good idea to practise what you will say for your showcase by speaking it out loud.

Showcasing is a way of sharing your project with other people.

Tips for showcasing your project:

- Speak loudly and clearly.
- Make sure that your audience knows which art piece is yours.

Showcase

Showcase your art piece. You should:
- state why you chose the colour palette that you did
- state how you think your design represents the school community.

Workbook page 49: Complete Task A, '**Reflection**'.

Project: Create a collaborative pixel art installation that represents your local community

Congratulations!

Well done! You have completed Chapter 4, 'How computers work'.

In this chapter you:

- ☑ explored automation in the health and advertising industries
- ☑ compared applications software with systems software
- ☑ discovered how binary is used to represent numbers, images, characters and sound
- ☑ created a collaborative piece of pixel art that represents your local community.

Key terms

Analogue sound – Sounds that are all around us in the non-digital world

Application software – Software used for a specific task, such as a word processor

Automation – Carrying out tasks with little to no human interaction

Characters – Numbers, letters and punctuation, for example, A, " or 9

Colour depth – Number of bits (1s and 0s) that are available for each pixel. The higher the number of bits, the more colours that are available

Decimal – Base-10 number system that uses 10 digits, from 0 to 9

Denary – Alternative for 'decimal'

Operating system – Software that manages the hardware, software and resources of a computer

Pixel – Block of colour that forms part of an image. A binary code is used to represent each colour

System software – Software that controls how the computer operates

Utility software – Software that provides additional support for a system, such as antivirus or disk cleanup management

Reflect: What was it like to see your piece of artwork as part of a collaborative installation? How well did your class collaboration represent your community?

Workbook page 49: Complete the '**Reflection**' task.

Chapter 5: Create with code 2

Project: Make a recommendation system

In this chapter, you will:

- write programs in Python, a text-based programming language
- plan and debug a Python program
- write a recommendation system in Python
- share and read a Python project.

End of chapter project: Python project

You will make a recommendation system that asks a user questions and then makes a recommendation based on their answers.

You will design something like these Python projects:

School club

```
What is your name? Farah
Hey Farah let's find the right club for you
What school year are you in?  7
Do you prefer art (a) or sport (s)   a
*******************************
I recommend  drawing club
*******************************
```

Activity recommendation

```
What is your name? Abi
OK Abi let's find you an activity
How many hours do you have?   3
Would you prefer indoors (i) or outdoors (o)? i
-~-~-~-~-~-~-~-~-~-~-~-~-~-~-~-
I recommend:  cinema
-~-~-~-~-~-~-~-~-~-~-~-~-~-~-~-
```

Book recommendation

```
What is your name?   Malik
Welcome to the book recommender Malik
How old are you?   10
Do you prefer robots (r) or elephants (e)?   e
~~~~~~~~~~~~~~~~~~~~~~~~~~~~~~~
I recommend The Girl Who Stole an Elephant
~~~~~~~~~~~~~~~~~~~~~~~~~~~~~~~
```

Project: Make a recommendation system

Chapter 5.1 Python: input and output

What do we already know?

- You can create computer programs by writing your own code.

What is Python?

A **text-based programming language** allows you to type text instructions that a computer can run.

Python is a popular text-based programming language that is used in industry. Python is popular for coding projects that involve lots of data or scientific calculations. It is also used for quick automation tasks.

Key terms

Text-based programming language – Allows you to type text instructions that a computer can run

Python – Text-based programming language that is popular in industry

Subroutine – Named section of code that can be called and performs a specific task

Function (Python) – Type of subroutine that returns a value. Python has built-in functions that are premade for you to use. You can also write your own functions in Python

Call (a subroutine) – To tell a program to execute the code within a subroutine

Argument – Input to a subroutine call

Statement – Complete instruction that the computer can run

What is EduBlocks?

EduBlocks is a programming tool that allows you to learn Python by dragging and dropping blocks. EduBlocks allows you to build on the skills you learnt from programming using Scratch.

Your teacher

 Your teacher will provide a handout that maps Scratch blocks to EduBlocks and Python.

Blocks in EduBlocks

EduBlocks blocks of Python code

the text-based version of the same blocks of Python code

Figure 5.1 You drag the blocks and EduBlocks creates the text-based Python code.

The EduBlocks editor allows you to drag the blocks of Python code that you need to create a program. You can then see those blocks of code as real Python code on the right-hand side of the screen.

> **Discuss 1**
>
> Can you spot the similarities between the code on the left and the code on the right? What are they?

> **Fact**
>
> EduBlocks was created by Joshua Lowe to help other children learn to use text-based programming. He started creating EduBlocks when he was 12 years old!

Use `print` to output text

In a previous stage, you created **subroutines** using the My Blocks feature of Scratch.

Figure 5.2 My Blocks in Scratch

In Python, a subroutine is called a **function**. Python has lots of built-in (premade) functions that you can use to do common tasks, such as printing some text on the screen.

Project: Make a recommendation system

To use a built-in function, you need to **call** it using its name, and provide at least one **argument**.

Here is a screenshot showing the built-in `print()` function. The word `print` is used to call the function. The text inside the brackets is the argument. The whole line of code is called a **statement**.

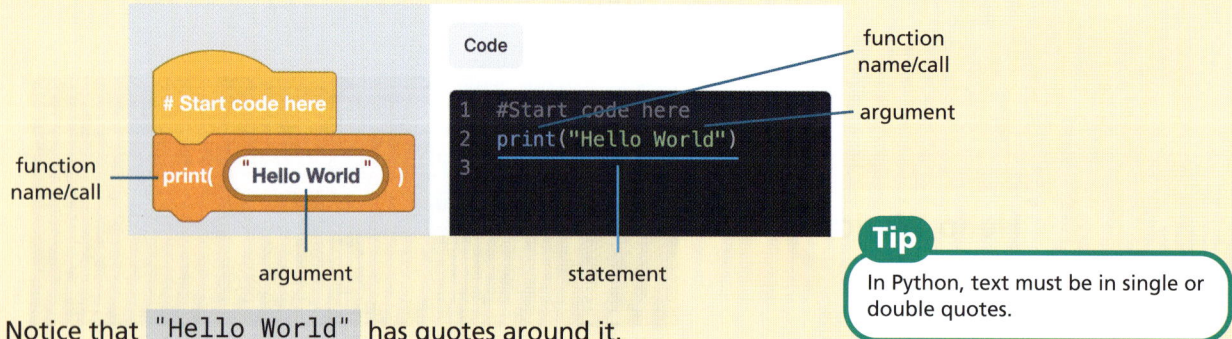

Notice that `"Hello World"` has quotes around it.

Tip

In Python, text must be in single or double quotes.

Discuss 2

When this Python code is run, what text do you think will be displayed on the screen as output? Why do you think this?

Your teacher

Your teacher will tell you how to access EduBlocks.

Build 1:

Write code using a print statement:

- Enter the code above into EduBlocks.
- Run your program.

Discuss 3

Was your prediction correct? How could you adapt the code to display a different message?

Build 2:

Adapt the code:

- Change the text from "Hello World" to text of your choice.
- Stop and Run your program.

Workbook page 50: Complete Task A, '**Print statements**'.

Chapter 5.1

Ask for input in Python

This program creates a variable called `name`. It then asks for input using the instruction "What is your name?". A user can then type their name and their answer will be assigned to the `name` variable.

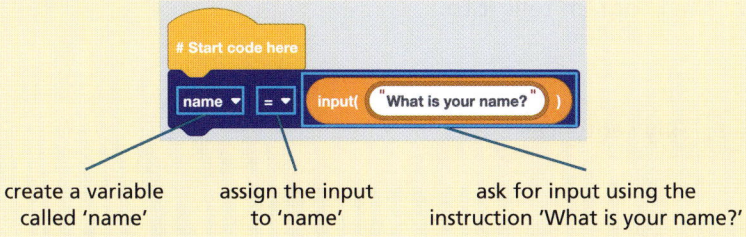

create a variable called 'name'   assign the input to 'name'   ask for input using the instruction 'What is your name?'

Discuss 4
What do you think will happen when this code is run? Why do you think that?

Build 3:
Write the code and run it to see what happens.

Use the print function to display a greeting

The program now has a new line of code. This code uses the `print()` function again. This time there are two arguments: `"Hello"` and `name`. Notice that the arguments are separated by a comma.

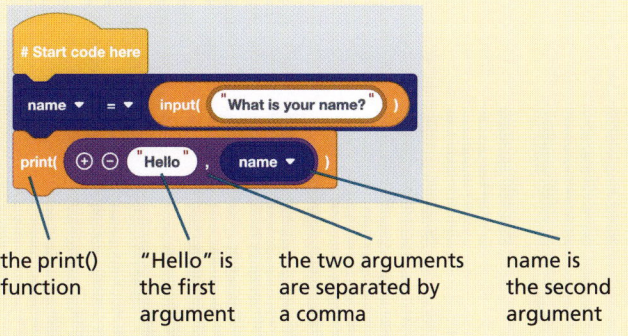

the print() function   "Hello" is the first argument   the two arguments are separated by a comma   name is the second argument

Tip
Notice that the purple 'text' block has been used inside the 'print()' statement.

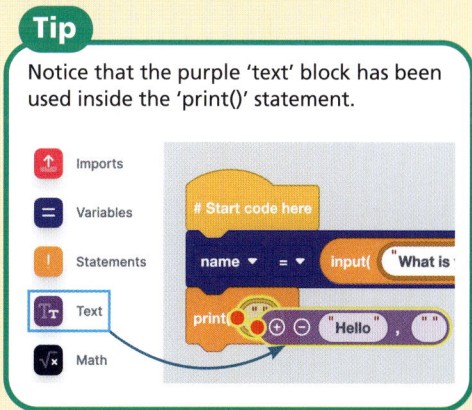

Discuss 5
What do you think will happen when this code is run? Why do you think that?

Build 4:
Write the code and run it to see what happens.

Tip
You can use Ctrl+Z to undo your last action in EduBlocks.

Project: Make a recommendation system

Silly sentences

> **Your teacher**
> Your teacher will demonstrate a silly sentence example that uses variables, input and print.

Notice that Python prints out *exactly* what the program tells it to. Even if it doesn't make sense.

Workbook pages 50 and 51: Complete Task B, '**Silly sentences**'.

Build 5: Write the code for your silly sentence.

Swap projects with a partner and try their project.

Reflect: In what ways is Python similar to Scratch? In what ways is it different?

Workbook page 52: Complete the '**Reflection**' task.

Chapter 5.2 Python: operators, conditions and data types

What do we already know?

- An arithmetic operator is an operator that performs arithmetic, such as add (+), subtract (-), multiply (*) and divide (/).
- A comparison operator is an operator that compares values and results as either True or False, such as less than (<), more than (>) and equal to (==).
- Debugging means finding errors in code and fixing them.

Key terms

Data type – Classification such as string or integer that tells the computer how to work with a value

Integer – Whole number

Arithmetic operators

Python has arithmetic operators for performing calculations using addition, subtraction, multiplication and division. In EduBlocks, these are available in the 'Math' section.

Data types

In Python, strings (text) and numbers are treated differently. This is because they each have a different **data type**. A data type, such as string or **integer** (whole number), tells the computer how to work with a value. The data type of a value makes a difference to the behaviour of operators.

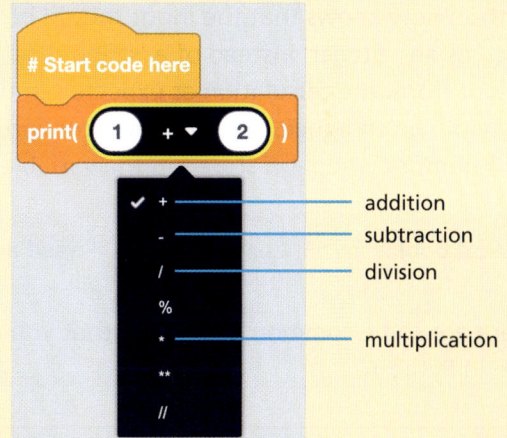

Workbook page 53: Complete Task A, '**Why do we need data types?**'.

Discuss 6

Can you explain the result you got?

The int() function for integers

In Python, we can tell the computer which data type we want it to use for a particular value.

The `input()` function returns a string. If we want the user input to be treated as an integer, then we use the `int()` function.

Project: Make a recommendation system

To modify the code to use the `int()` function:

1. Drag the `input("amount:")` blocks to one side for a moment.

2. Find the `int()` function in the 'Statements' menu and drag it into position.

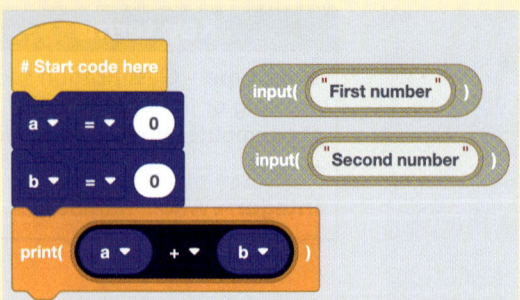
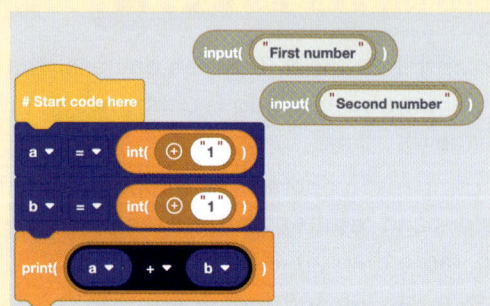

3. Next, drag the two `input()` statements inside the `int()` block.

Python now knows that the input should be used as an 'integer' instead of a 'string' and will perform the calculation as expected. In the example given above, the amount is converted to a number when it is stored in the variable.

Workbook page 53: Complete Task B, '**I need an int**'.

Here are some common data types that you can use in Python.

Function	Data type	Description	Example
int()	Integer	Positive and negative whole numbers	2, 16, −1, 0
float()	Float (Floating point number)	Decimal numbers that can have a fractional part after the decimal point	4.5, −5.6, 3.14159
str()	String	Strings of text made up from characters, written inside double or single quotes.	`"Hello world"` `'total:'` `"Amazing 🎉"` (emoji can appear in Strings)
bool()	Boolean	True or False values used in conditional logic	True, False

Table 5.1 Data types in Python

Workbook page 54: Complete Task C, '**Using decimal numbers**'.

Discuss 7

What happened when you ran the code? Why do you think that?

Chapter 5 Create with code 2

If you try to pass a number with a decimal point into `int()` you will get an error message. The important part is 'invalid literal for `int()`'. Python is telling you that it can't convert the input into an integer. You will need to use a different data type. The error message can help you to debug your program.

Use the float() function for decimal numbers

The code (right) has been created to calculate the total ticket price for a group of people. It is showing an error message when the code is running because the ticket price is a decimal number.

If you want to perform calculations on decimal numbers in Python, then you will need to use the 'float()' function.

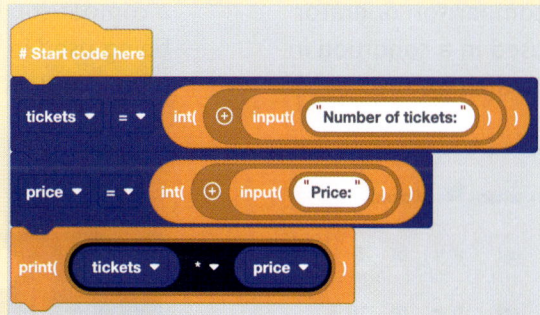

To swap an `int()` to a `float()` in EduBlocks you:

1. Drag the `input()` block to one side.

2. Next, remove the `int()` block by dragging it over to the left (or select it and press delete).

3. Now, find the `float()` block in 'Statements' and drag it in place.

4. Then you can add the `input()` block back in.

Workbook page 54: Complete Task D, '**Calculate ticket prices**'.

Comparison operators in Python

Comparison operators in Python allow you to compare values. The table shows the Python comparison operators and the operations that they perform.

When a comparison operator is used inside a condition, it returns either `True` or `False`.

Operator	Operation
<	Less than
>	More than
<=	Less than or equal to
>=	More than or equal to
==	Equal to
!=	Not equal to

Table 5.2 Python comparison operators

You will have seen a comparison operator used in a condition in Scratch before.

Here is an example of a comparison operator being used in Python with EduBlocks.

This example creates a variable called 'a' and assigns it the number '5'. It then creates a variable called 'b' and assigns it the number '10'. There is then a condition using an if statement to check if 'a' is less than 'b'. If that is True then it will output 'a is less than b'.

> **Build 6:**
>
> Create the example program above in EduBlocks to practise using an 'if' block and a comparison operator. Try changing the numbers assigned to 'a' and 'b' to see if the output changes.

Workbook page 54: Complete Task E, '**True or False?**'.

> **Reflect:** Why are data types important in Python? What data types do you know and what values can they have?

Workbook page 55: Complete the '**Reflection**' task.

Chapter 5.3 Python: Boolean logic, comments and functions

What do we already know?

- Boolean logic is a form of algebra that uses the operators AND, OR and NOT.
- A comment is a note written for a human to read in a program. All comments are completely ignored by the program.
- A subroutine is a named section of code that can be called and performs a specific task.

Key terms

Parameter – Special named variable that is part of a subroutine definition

Boolean logic

Python uses the Boolean operators AND, OR and NOT. You can use these to combine conditions using Boolean logic.

This example asks the user for two input values. It uses Boolean logic to make a book recommendation.

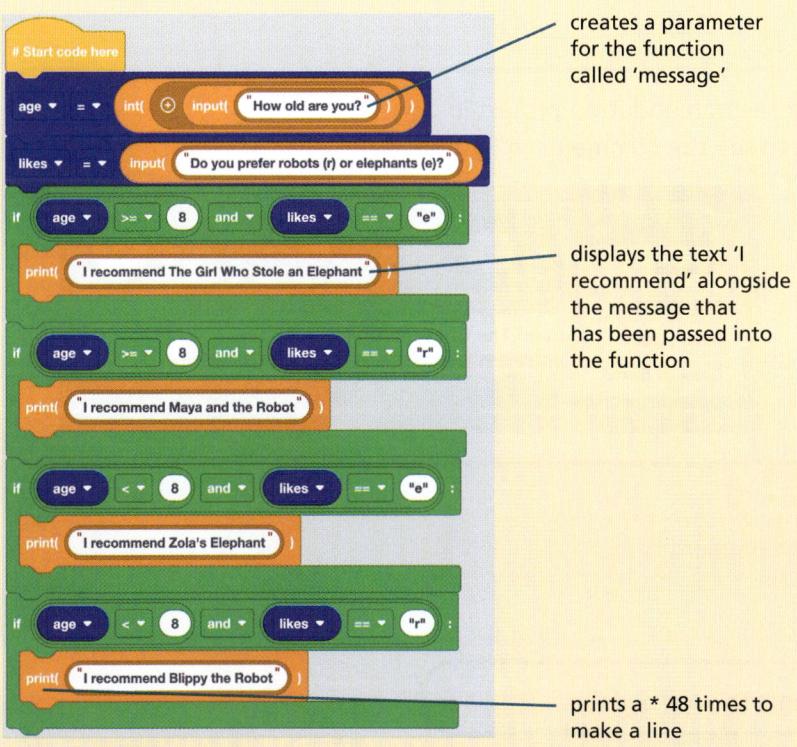

creates a parameter for the function called 'message'

displays the text 'I recommend' alongside the message that has been passed into the function

prints a * 48 times to make a line

Your teacher

Your teacher will share a copy of this project with you.

Workbook page 56: Complete Task A, '**Decisions**'.

Project: Make a recommendation system 75

Comments

A comment is a note written for a human to read in a program. All comments are completely ignored by the program.

In Python, a comment begins with the '#' character.

This is the Python code for the book recommendation example with comments added.

> **Tip**
> You say 'hash' when you say it out loud.

```
Code
 1  #Start code here
 2  age = int(input("How old are you? "))
 3  likes = input("Do you prefer robots (r) or elephants (e)? ")
 4  # Older child who likes elephants
 5  if age >= 8 and likes == "e":
 6      print("I recommend The Girl Who Stole an Elephant")
 7  # Older child who likes robots
 8  if age >= 8 and likes == "r":
 9      print("I recommend Maya and the Robot")
10  # Younger child who likes elephants
11  if age < 8 and likes == "e":
12      print("I recommend Zola's Elephant")
13  # Younger child who likes robots
14  if age < 8 and likes == "r":
15      print("I recommend Blippy the Robot")
16
```

Build 7:

Add helpful comments to the book recommendation project in EduBlocks. Check the Code output to see the comments in Python.

To add a comment in EduBlocks:

- Right-click on the block you want to add a comment to.
- Select 'Add Comment'.
- Click on the comment icon.
- Type or edit your comment.

Workbook page 56: Complete Task B, '**Comments**'.

Discuss 8

Why is it important to add comments to your code?

Chapter 5.3

Define your own functions

Remember that `print()` and `input()` are built-in functions in Python. You can define your own functions in Python as well. This is similar to 'My Blocks' in Scratch.

This is useful when you have groups of statements that are repeated in multiple places in your code.

> **Workbook** page 57: Complete Task C, '**Recognising patterns**'.

Create a function to avoid repeating code

By spotting patterns, you can see where a function might be useful to help you to avoid repeating lines of code.

The three lines of code have now been placed inside a function called 'show_message'.

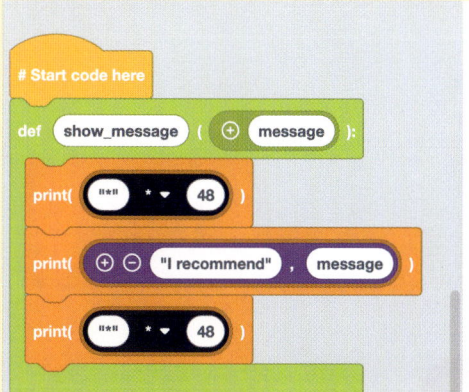

Tip

Variable and function names in Python use the '_' (underscore) character to separate words to make them easy to read. This is called 'snake case' because the name goes up and down like a snake!

This function can now be called whenever you want to display a recommendation.

The example below shows the function being called.

call the function — show_message ("The Girl Who Stole an Elephant") — pass the message "The Girl Who Stole an Elephant"

When the complete code is run, the output will be:

```
************************************************
I recommend The Girl Who Stole an Elephant
************************************************
```

The '*' symbol has been printed 48 times above and below the recommendation. The message displays the book recommendation.

Project: Make a recommendation system — 77

The function call 'show_message' can now be used within the if statement to avoid repeating the three lines of code.

if the age is more than or equal to 8 and they like elephants

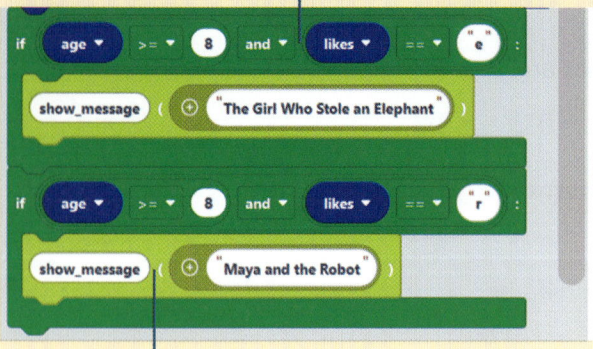

call this function recommending the elephant book

> **Workbook** pages 57 and 58: Complete Task D, '**Modify the recommendation app**'.

Debug a function

When you create your own functions it is important to test them to see if there are any errors. If you spot an error (bug) then you should try to fix it (debug).

> **Workbook** pages 58 and 59: Complete Task E, '**Debugging a function call**'.

> **Reflect:** Did you have any errors in the 'Modify the recommendation app' task? If so, how did you fix them?

> **Workbook** page 59: Complete the '**Reflection**' task.

Chapter 5.4 Start your recommendation project

What do we already know?

- How to write Python projects in EduBlocks using input, output, operators, functions and comments.

Project brief

You are going to create a recommendation system that takes inputs from the user and makes a recommendation. This will be similar to the one that you explored in Lesson 3.

Your project should:
- greet the user using their name
- use at least one text input and at least one numeric input (integer or float)
- use conditional operators
- define a function and call it multiple times
- have helpful comments
- use meaningful variable names.

School club

```
What is your name? Farah
Hey Farah let's find the right club for you
What school year are you in?  7
Do you prefer art (a) or sport (s)   a
*******************************
I recommend  drawing club
*******************************
```

Activity recommendation

```
What is your name? Abi
OK Abi let's find you an activity
How many hours do you have?  3
Would you prefer indoors (i) or outdoors (o)? i
-~-~-~-~-~-~-~-~-~-~-~-~-~-~-~-
I recommend:  cinema
-~-~-~-~-~-~-~-~-~-~-~-~-~-~-~-
```

Book recommendation

```
What is your name?  Malik
Welcome to the book recommender Malik
How old are you?  10
Do you prefer robots (r) or elephants (e)?   e
~~~~~~~~~~~~~~~~~~~~~~~~~~~~~~~~
I recommend The Girl Who Stole an Elephant
~~~~~~~~~~~~~~~~~~~~~~~~~~~~~~~~
```

Project: Make a recommendation system

A complete example

This is a complete example of a finished recommendation system project. You can use this for reference as you work on your project.

Plan your recommendation system

Think about what kind of recommendation system you want to make. You could choose books, as in the example, or perhaps something you're an expert on. For example: places to visit, activities where you live, computer games, films (movies), sports, recipes, school clubs, hobbies, pets or something else.

You will need to greet the user by name to make the recommendation feel personal. (Have you noticed how some apps do this?)

You will need to choose <u>one string input</u> and <u>one number input</u> (integer or float) to get data to use for your recommendation.

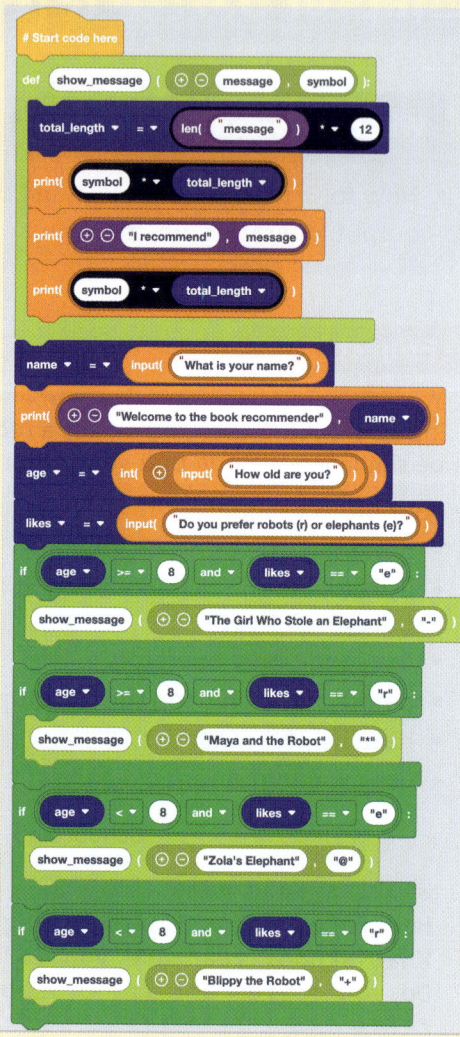

> **Workbook** page 60: Complete Task A, '**Plan your recommendation system**'.

> **Your teacher**
> Your teacher will provide a starter project that provides some of the code you need.

Build 8:

When you have completed your plan, start working on your project. Remember to clone the example so you can work on it. You will have time to complete your project in the next lesson.

Tasks to complete this lesson:

- Create the greeting.
- Get the user input.
- Make decisions on which recommendation to make.

Save your project. Or take a screenshot of your code so you can build it quickly next lesson if you don't have an account.

Tip
The starter project has the 'age' and 'likes' variables to support you with the code structure. You can create your own variables to match your design and select them using the drop-down menu.

Tip
To support you with debugging your code, the if statements have been disabled to allow you to run parts of your code and test as you go. You can enable the code again by right-clicking on the block and choosing 'Enable code'.

Discuss 9

What errors have you found in your code and how did you fix them?

Tip
Give your project a name that matches the theme of your project. You can do this by going to 'Settings' in EduBlocks.

> **Workbook** page 61: Complete the '**Reflection**' task.

Chapter 5.5 Complete your recommendation project

> **What do we already know?**
> - How to use EduBlocks to develop text-based programs that use input and outputs.

Complete your recommendation system

> **Project brief**
>
> You are going to create a recommendation system that takes inputs from the user and makes a recommendation.
>
> Your project should:
> - greet the user using their name
> - use at least one text input and at least one numeric input (integer or float)
> - use conditional operators
> - define a function and call it multiple times
> - have helpful comments
> - use meaningful variable names.

Steps for completing your project

Below is a reminder of the steps that you will need to take to complete this project.

Project checklist:

1. Create the greeting.
2. Get the user input.
3. Make decisions (complete the if statements) on which recommendation to make.
4. Format the output nicely.
5. Add comments.
6. Test your code.
7. Improve your code.

Use the Project checklist in your Workbook to track the tasks that you have finished.

Workbook page 62: Complete Task A, '**Project checklist**'.

> **Discuss 10**
> How much progress have you made?

As you work through the tasks, check that you are meeting the project brief using the 'Meeting the brief' checklist.

Workbook page 62: Complete Task B, '**Meeting the brief**'.

Test your project

You should always test that your program works as expected to make sure that it meets the project brief. Use the test table in the 'Testing' task to check that your program works as expected.

Workbook page 63: Complete Task C, '**Testing**'.

Discuss 11

What are you looking forward to about sharing your project? Are you concerned about any parts of your project?

Workbook page 63: Complete the '**Reflection**' task.

Chapter 5.6 Share your recommendation project

What do we already know?

- A showcase is a time to share what you have created with an audience.
- Practising what you will say for your showcase is a good idea.

Sharing your project

Stay safe
 Make sure your project doesn't contain any personal information before you share it.

Showcase
Share your EduBlocks project so that anyone with a link can access it and try it out.

Your teacher will tell you if you will be sharing your projects on EduBlocks or inviting someone to your computer to share your project with them.

Try it out!

Try out the project you are given and give helpful feedback in the programmer's Workbook.

Workbook page 64: Complete Task A, '**Feedback**'.

Feedback

Work with other students to improve your project, based on feedback.

Workbook pages 64 and 65: Complete Task B, '**Reflection**'.

Discuss 12
When you tested a partner's program, what did you think of the recommendations you were given? Do you think you will follow the recommendations?

Congratulations!

Well done! You have completed Chapter 5, Create with code 2.

In this chapter you:

- ☑ wrote programs in Python, a text-based programming language
- ☑ planned and debugged a Python program
- ☑ wrote a recommendation system in Python
- ☑ shared and read a Python project.

Key terms

Argument – Input to a subroutine call

Call (a subroutine) – To tell the program to execute the code within a subroutine

Data type – Classification such as string or integer that tells the computer how to work with a value

Function (Python) – Type of subroutine that returns a value. Python has built-in functions that are premade for you to use. You can also write your own functions in Python

Integer – Whole number

Parameter – Special named variable that is part of a subroutine definition

Python – Text-based programming language that is popular in industry

Statement – Complete instruction that the computer can run

Subroutine – Named section of code that can be called and performs a specific task

Text-based programming language – Allows you to type text instructions that a computer can run

Reflect: Discuss: How did you enjoy creating a recommendation system that could be used by others?

Workbook page 65: Complete the '**Reflection**' task.

Chapter 6: Connect the world

Project: Create a method for encrypting text

In this chapter, you will:

- explain how IP, DNS and HTTPS standards help to create a reliable and secure internet
- compare different wireless communication technologies and explain how they are used
- create your own encryption algorithm for a substitution cipher
- crack an encryption algorithm by decoding a message, and relate your experience to modern internet security.

End of chapter project: Encryption algorithm

You will design an encryption algorithm and use it to encrypt messages.

Your encrypted messages could look something like these examples:

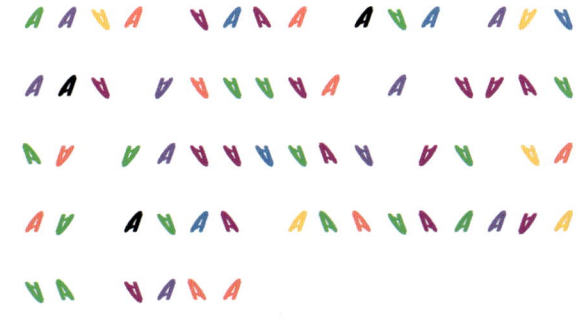

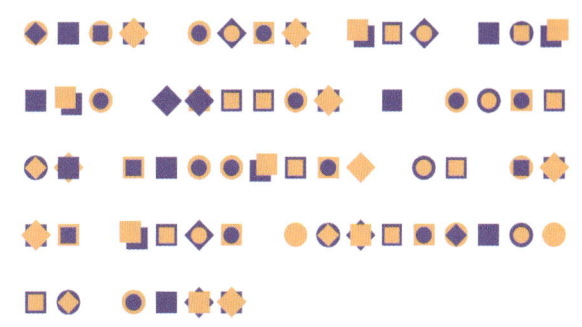

Project: Create a method for encrypting text 85

Chapter 6.1 Internet standards: DNS and HTTPS

What do we already know?

- All devices on a network have an IP address that is used to identify the destination of data.
- Web pages have addresses known as URLs.
- Data needs to be kept secure during transmission.

Key terms

Domain name – Human-readable label that is assigned to an IP address

Hypertext Transfer Protocol Secure (HTTPS) – Secure way of sending data between a web browser and a website

Top-level domain (TLD) – The last part of the domain. For example .com

Domain Name System (DNS) – Naming system and way of mapping from human-readable domain names to IP addresses

The parts of a URL

A Uniform Resource Locator (URL), or web address, is made up of several parts:

http:// www.example.com
↳Scheme ↳Domain Name

The scheme specifies the protocol or standard that the web browser uses to communicate with the website.

The **domain name** is a human-readable label for a website that a person can remember and type.

HTTP vs HTTPS

URLs for web pages use either the scheme HTTP or HTTPS.

In the early days of the World Wide Web (WWW), all websites used Hypertext Transfer Protocol (HTTP). HTTP is not secure so other devices on a network may be able to read data sent using HTTP. **Hypertext Transfer Protocol Secure (HTTPS)** is a secure version of HTTP that encrypts data before it is sent. Most popular websites now use HTTPS.

Stay safe

 You should always make sure a website is secure before sending personal data.

To check that a website is secure, look for an icon to the left of the URL in your web browser. Click the icon to see the settings for the website. If the website connection is secure (using HTTPS), then you will see a message such as 'This connection is secure'.

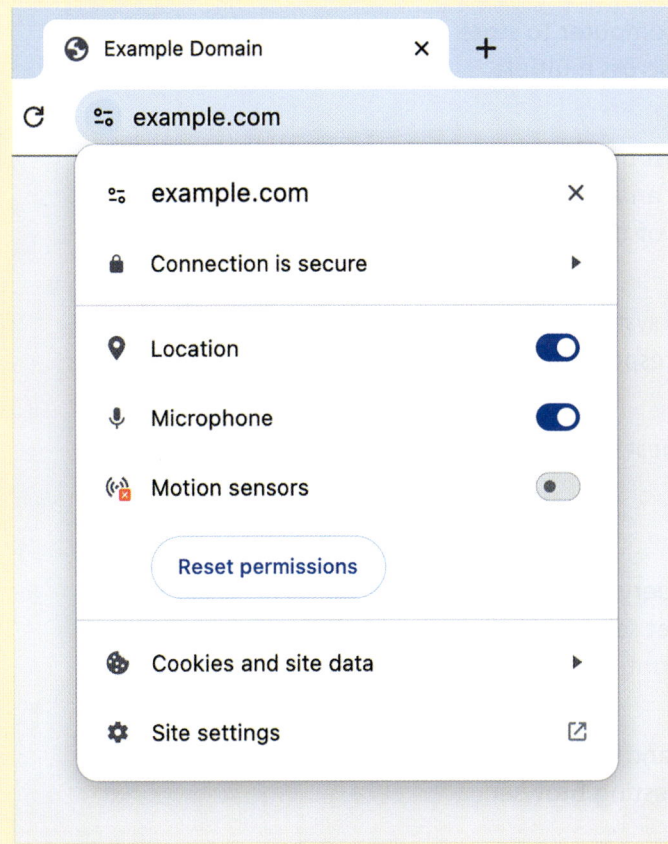

Workbook page 66: Complete Task A, '**Identify the parts of a URL**'.

Discuss 1
Why is website security important?

Domain names

The last part of a domain is the **top-level domain (TLD)**. TLDs can be generic, for example .edu, .com and .org, or they can be country specific, for example .fr (France), .kw (Kuwait), .in (India) and .so (Somalia). There are over 1500 TLDs! Newer TLDs include those for industries, for example .engineering and .software, and brands such as .microsoft and .cisco.

The main domain is the part before the TLD, such as 'example' in 'example.com'. Sometimes you will see a subdomain before the main domain, such as 'images' in 'images.google.fr'.

Workbook page 66: Complete Task B, '**Top-level domains**'.

Domain Name System (DNS)

Although domain names are friendly for humans to read and remember, computers need IP address numbers to send data to their destination.

There are far too many domain names for one computer to store a complete list. Therefore the list is distributed across multiple servers and systems.

When someone gets a new domain name, it must be registered with a domain registry. The domain registry has an authoritative DNS server that has a record of the IP addresses of the domains that it is responsible for.

The **Domain Name System (DNS)** is a naming system that allows your web browser to look up the IP address corresponding to a domain name.

When you visit a web page, how does your request get to the server with that web page?

Follow these steps in the diagram to find out.

1–2 Your web browser sends a request, a DNS query, to a DNS server. This is often administered by your internet service provider. The DNS server responds with the address of a root DNS server (there are around 13 of these in the world).

3–4 The requested URL is looked up in the DNS and the IP address of the website's location is returned to the requesting browser.

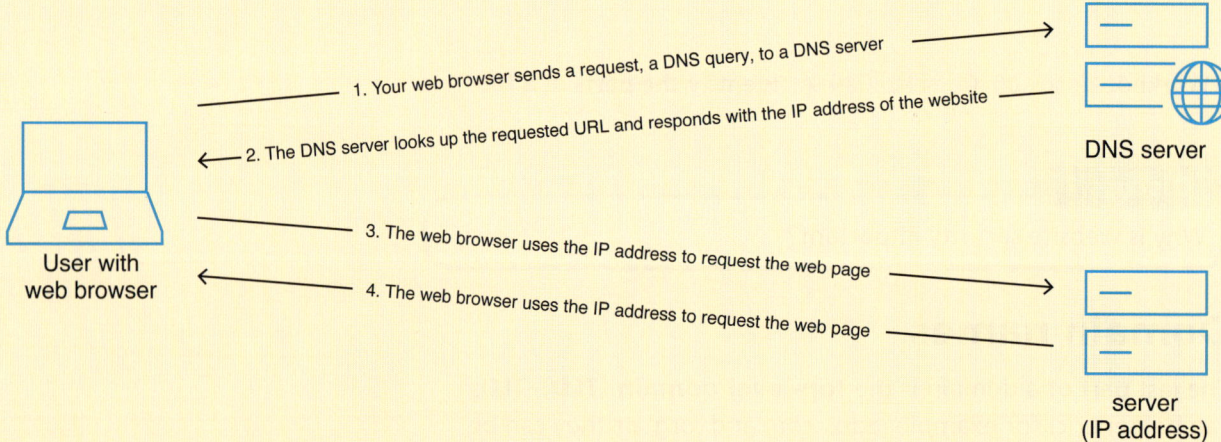

Chapter 6.1

Your web browser and the DNS servers can store a copy of an IP address they have requested so that it's quicker to respond in future. They can store a copy for only a limited period of time because IP addresses can change.

A DNS lookup is distributed across multiple computers so that no single computer needs to know the IP address for every domain name in the world.

> **Discuss 2**
>
> In pairs, follow the diagram, explaining the steps to each other. What role does the Domain Name System play in accessing a website from a web browser?

Workbook page 66: Complete Task C, '**The Domain Name System**'.

DNS security

A DNS is not secure. This is a security threat as you could be directed to the wrong website or an attacker could get access to your data. A new standard called DNS over HTTPS (DoH) allows DNS lookups to be made securely.

> **Reflect:** Why is it important to have standards such as IP, DNS and HTTPS?

Workbook page 67: Complete the '**Reflection**' task.

Project: Create a method for encrypting text

Chapter 6.2   Wireless data transmission

> **What do we already know?**
> - Digital devices can transfer data wirelessly using radio waves, including Wi-Fi and cellular networks.
> - It's important to keep data secure during transmission.
> - The Internet of Things (IoT) is a network of everyday devices that are connected through the internet.

> **Key terms**
>
> **Bluetooth** – Wireless standard for communicating between nearby devices
>
> **4G and 5G** – Versions of cellular network technology with improvements over previous generations
>
> **Packet loss** – When 'packets' of data fail to reach their destination
>
> **Bit errors** – When data gets changed during transmission

Wireless communication

Bluetooth, Wi-Fi and cellular networks all provide ways for devices to communicate wirelessly. They have different properties to support different situations.

	Bluetooth	Wi-Fi	Cellular network
Distance (range)	Nearby devices	Usually in the same building	Local neighbourhood
Data transfer speed	Low	Medium–high	Medium–high
Hardware cost	Low	Medium	High
Data transmission cost	None	Medium – part of internet package	High – part of mobile package
Power consumption	Low	Medium–high	Medium–high
Security	Low–medium	Medium–high	High
Impact of interference	High	Medium	Low
Network	None	Internet	Cellular network
Infrastructure	None	Wireless access points connected to the internet	Base stations in cellular masts connected to a mobile service provider's network

Workbook page 68: Complete Task A, '**Which wireless technology?**'.

> **Discuss 3**
>
> Why do we have multiple wireless communication technologies? Couldn't we just have one?

Chapter 6.2

Cellular networks

Base stations

Mobile phones communicate with each other by connecting to a local base station. Base stations are inside cell towers. The base stations can be connected to each other using wired or wireless connections, including communication via satellite.

Mobile phones need to connect to a base station for their mobile provider. Some cell towers contain more than one base station.

A cell is the area covered by a single base station in a cell tower.

> **Discuss 4**
>
> Why do you think mobile providers share cell towers?

4G, 5G and 6G

4G, **5G** and 6G refer to versions of the cellular network standard.

4G is widely used and there is increasing support for 5G around the world. 5G improvements include the following:

- significantly faster data transfer speeds
- increased network capacity, which is important for densely-populated areas and more Internet of Things devices
- smaller local cells in urban areas
- improved energy efficiency
- faster reliable communication for devices.

What about 6G? 6G is already being planned and will offer further improvements including supporting even more Internet of Things devices. 6G is expected to launch in the 2030s.

Workbook page 68: Complete Task B, '**5G features**'.

Project: Create a method for encrypting text

Errors in wireless transmission

All forms of wireless communication using radio waves can suffer from errors.

Packet loss occurs when 'packets' of data fail to reach their destination.

Bit errors occur when data gets changed during transmission.

Causes of errors include:

- interference due to electrical 'noise'
- interferences from other devices using the same radio frequency or from reflection of signals
- poor connection due to physical obstacles, especially metal objects
- network collisions when devices on the same network try to send data at exactly the same time.

Cellular signals:

- are better at passing through physical obstacles than Bluetooth or Wi-Fi
- experience less interference from other electrical devices
- can experience errors when switching between cells and moving a connection from one base station to another.

Computer systems cannot guarantee that data will be transmitted without error so they must be designed to detect and handle errors.

> **Reflect:** How can you reduce the chance of errors when using wireless transmission?

Workbook page 69: Complete the '**Reflection**' task.

Chapter 6.3 — Encryption and data security

What do we already know?

- Where and why encryption is used in digital systems.
- How to write and decode messages using the Caesar cipher and the Pigpen cipher.

Key terms

Encryption algorithm – Specific set of rules for turning a plaintext message into encrypted ciphertext, sometimes with the use of a key

Decryption algorithm – Specific set of rules for turning a ciphertext message into a plaintext message, sometimes with the use of a key

Key – Piece of information that can be used with a decryption algorithm to decrypt a ciphertext message

Substitution cipher – Way of encoding messages where a letter is replaced by another letter or symbol

Encryption

Encryption is a way of converting information or data so that it can't easily be read during transmission and only the intended recipient can access the original data.

An **encryption algorithm**, or cipher, is a specific set of rules for turning a plaintext message into encrypted ciphertext, sometimes with the use of a key.

A **decryption algorithm** is a specific set of rules for turning a ciphertext message into a plaintext message, sometimes with the use of a **key.**

The Caesar cipher encodes messages by replacing each letter with another that is a particular number of places forwards or backwards in the alphabet. The number of places forwards or backwards is called the shift. The shift is the key.

```
SFNFNCFS    EP    OPU    S
FQMZ    UP    USPMMT    CF
DBVTF    JU    KVTU    NBL
FT    UIFN    XPSTF
```

This message is encrypted with a forward shift of 1.

Project: Create a method for encrypting text

The Pigpen cipher encodes messages by replacing each letter with a symbol based on a grid. Pigpen is designed so that it's easy to remember the key.

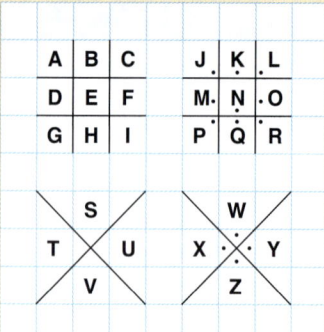

Here is an example of a message encoded using the Pigpen cipher:

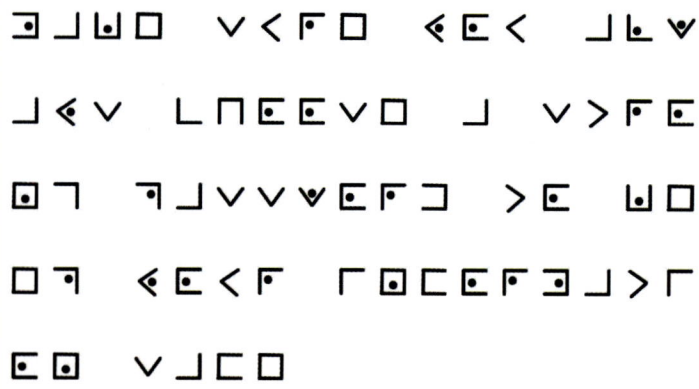

Workbook page 70: Complete Task A, '**Decryption**'.

Data security

Encryption plays an important part in keeping your data safe online.

When your web browser uses HTTPS to communicate with a website, your data is encrypted using a secure encryption algorithm (sometimes with the use of a key). This means that if someone gets access to your data they will not be able to read it.

HTTPS uses algorithms that are much more secure than the Pigpen or the Caesar cipher, but the principles are the same. Data is encrypted before transmission and then decrypted by the recipient using an agreed key. If someone intercepts your data on the network, they will not have access to the key, so they won't be able to understand the data.

Workbook page 70: Complete Task B, '**Encryption**'.

What is a substitution cipher?

A **substitution cipher** is a way of encoding messages by replacing a letter by another letter or symbol. Caesar and Pigpen are both substitution ciphers.

One of the oldest known substitution ciphers is Atbash. To use Atbash, you swap pairs of letters starting with the first and last letters, and then the second and second to last letter, and so on.

Atbash was originally designed for the Hebrew alphabet, but it can be adapted for other alphabets. Atbash is named after Hebrew letters (Aleph–Taw–Bet–Shin) at the beginning and end of the alphabet.

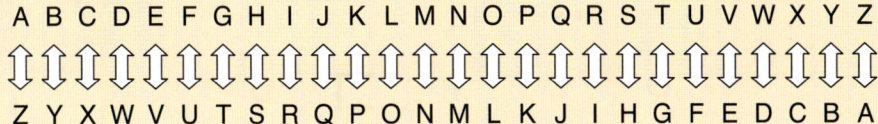

Here is an example of a message coded using the Atbash cipher:

```
Plaintext.  THIS IS A MESSAGE IN ATBASH
Ciphertext. GSRH RH Z NVHHZTV RM ZGYZHS
```

> **Discuss 5**
>
> What is the decryption algorithm for Atbash? How easy would it be to decrypt a message coded using Atbash? How does it compare with the Pigpen and Caesar ciphers?

Encrypt a message

Your teacher will give you a link to a Scratch project that can encrypt a message using either Atbash, the Caesar cipher or the Pigpen cipher. You are going to encrypt a message and then swap with a partner who will decrypt the message.

The message to your partner should have 4–5 words encrypted using either the Pigpen cipher, Atbash or the Caesar cipher.

To encrypt your message:

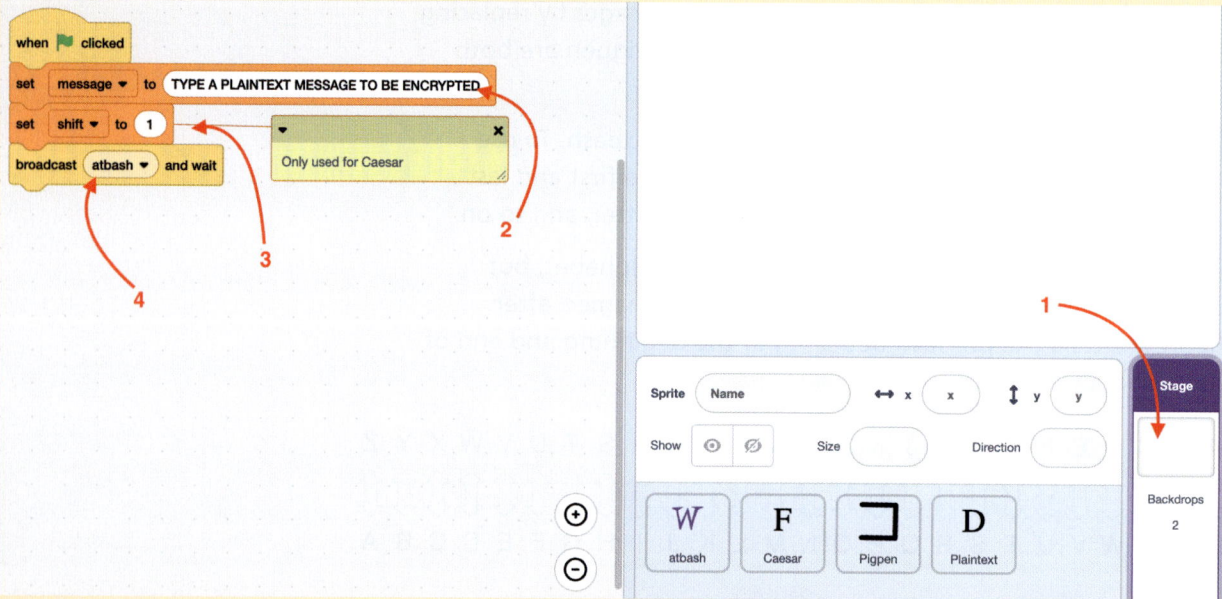

- Open the Scratch project.
- Click 'See inside'.
- Click on the Stage code area.
- Type your message into the (message) variable.
- If you are using the Caesar cipher, type the shift number into the (shift) variable. This should be a number from 1 to 25. Alternatively, you could use a negative number to shift in the reverse direction.
- Select the message for your chosen encryption algorithm in the 'broadcast' block.

Click the green flag to run the encryption algorithm and create ciphertext.

Now swap Workbooks with a partner.

Workbook page 70: Complete Task C, '**Decrypt a message**'.

Workbook page 71: Complete Task D, '**Explore the encryption algorithm**'.

Reflect: How much more difficult would it be to decrypt a message if you didn't know the encryption algorithm and didn't have the key?

Workbook page 71: Complete the '**Reflection**' task.

Chapter 6.4 — Create your own encryption algorithm

What do we already know?

- An encryption algorithm is a specific set of rules for turning a plaintext message into encrypted ciphertext.
- A substitution cipher is a way of encoding messages whereby a letter is replaced by another letter or symbol.

Project brief

You will design a method for encrypting text. You will create a substitution cipher that replaces each letter of the alphabet with a different letter or symbol.

You will need to choose a letter or symbol to substitute for each letter of the 26 letters of the alphabet. This will be the key for your encryption algorithm.

The same message has been encrypted with three different substitution ciphers that are similar to the ones you will make.

Emoji cipher – This cipher uses a different emoji character for each letter of the alphabet.

AAA Text and colour cipher – This cipher uses the letter A in the Scratch Marker font with different colours and rotations created in the Scratch Paint editor.

Shapes – This cipher uses unique shapes created in the Scratch Paint editor. This is an example of a code that could be turned into a fabric design and hidden in plain sight!

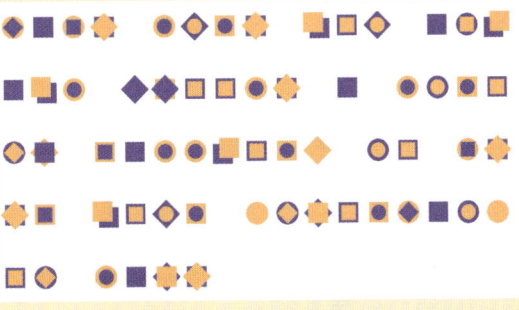

Project: Create a method for encrypting text

Plan your encryption algorithm

Your teacher will demonstrate how your encryption algorithm will work.

Workbook page 72: Complete Task A, '**Plan your substitution cipher**'.

Your teacher
Your teacher will provide you with a link to a Scratch starter project with the code for your encryption algorithm.

Start making your encryption algorithm

Build 1:
Add costumes to the MyCipher sprite to start making the key for your encryption algorithm.

Tip
You can click the green flag at any point to test your encryption algorithm.

Workbook page 72: Complete Task B, '**Start creating your substitution cipher**'.

Save
Save your Scratch project.

Workbook page 73: Complete Task C, '**Save and load your Scratch project**'.

Reflect: Discuss: Did you overcome any issues with your Scratch project? If so, share them.

Workbook page 73: Complete the '**Reflection**' task.

Chapter 6.5 Complete your encryption algorithm

What do we already know?

- A decryption algorithm is a specific set of rules for turning a ciphertext message into a plaintext message, sometimes with the use of a key.
- A key is a piece of information that can be used with a decryption algorithm to decrypt a ciphertext message.

Key terms

Crack – To find out how to get the plaintext from encrypted ciphertext without knowledge of the decryption algorithm or key

Frequency analysis – Knowledge about how often each letter is used in text written in a particular language

Cracking a code

To **crack** a code means finding out how to get the plaintext from encrypted ciphertext without knowledge of the decryption algorithm or key.

One way of cracking a code is to use 'brute force', which means trying all the different combinations until one makes sense. This takes a very long time for a person using a substitution cipher, but a modern computer can do this very quickly.

Another way of cracking a code is 'social engineering' where you try to get a person to give you a clue accidentally.

Codes are sometimes cracked when the same word appears frequently and so it can be guessed. This could be a person's name or a place. Once some letters are known, it is easier to crack the rest of the code.

Frequency analysis

A common way to crack substitution ciphers is **frequency analysis**. Frequency analysis uses knowledge about how often each letter of the alphabet is used in text written in a particular language.

The first person known to have used frequency analysis is al-Kindi, a mathematician and philosopher who was alive around 801–873 CE. Al-Kindi was born in Kufa in Iraq.

Project: Create a method for encrypting text

Al-Kindi wrote a book called *Manuscript on Deciphering Cryptographic Messages*. Frequency analysis is one of the techniques described in the book for deciphering a message or for cracking a code.

Complete your code

> **Build 2:**
>
> Add more sprite costumes to complete your code.
>
> Prepare a message and clues.
>
> 1. Add a message of your choice to the 'message' variable in the code for the Stage.
> 2. Edit the backdrop to add your team name to identify your code.

"One way to solve an encrypted message, if we know its language, is to find a different plaintext of the same language long enough to fill one sheet or so, and then we count the occurrences of each letter. We call the most frequently occurring letter the 'first', the next most occurring letter the 'second', the following most occurring the 'third', and so on, until we account for all the different letters in the plaintext sample".

"Then we look at the cipher text we want to solve and we also classify its symbols. We find the most occurring symbol and change it to the form of the 'first' letter of the plaintext sample, the next most common symbol is changed to the form of the 'second' letter, and so on, until we account for all symbols of the cryptogram we want to solve"

Workbook page 74: Complete Task A, '**Prepare a message**'.

Workbook page 74: Complete Task B, '**Prepare a clue**'.

> **Build 3:**
>
> Right-click on the Scratch stage and choose 'Save image as...' to save your ciphertext message as an image.
>
> Send a copy of your ciphertext message to a printer and print it out.

Workbook page 74: Complete Task C, '**Cracking your code**'.

> **Reflect:** Have you ever cracked a code without knowing the encryption algorithm or key? How did you do it?

Workbook page 75: Complete the '**Reflection**' task.

Chapter 6.6 Cracking the code

> **What do we already know?**
> - Frequency analysis is knowledge about how often each letter is used in text written in a particular language.

Letter frequency

Letter frequencies in written text in the English language are roughly as shown on the right.

Letter	Frequency
A	8.2%
B	1.5%
C	2.8%
D	4.3%
E	12.7%
F	2.2%
G	2.0%
H	6.1%
I	7.0%
J	0.15%
K	0.77%
L	4.0%
M	2.4%
N	6.7%
O	7.5%
P	1.9%
Q	0.095%
R	6.0%
S	6.3%
T	9.1%
U	2.8%
V	0.98%
W	2.4%
X	0.15%
Y	2.0%
Z	0.074%

Cracking the code

> **Showcase**
> Your teacher will give you a printed code from another pair. You will need to crack their code.

Use the space in your Workbook to make notes as you try to decrypt the ciphertext and reveal the plaintext.

Workbook page 76: Complete Task A, '**Crack the code**'.

Workbook page 76: Complete Task B, '**Clue**'.

Write down the plaintext message in your Workbook when you have decrypted it.

Workbook page 77: Complete Task C, '**Plaintext message**'.

> **Your teacher**
> Your teacher will tell you when to complete the 'I cracked a code!' task.

Workbook page 77: Complete Task D, '**I cracked a code!**'.

> **Tip**
> Think about which letters can be double letters in a word.

> **Discuss 6**
> What helped you to crack the code? What made it more difficult?

Workbook pages 77 and 78: Complete Task E, '**Reflection**'.

Project: Create a method for encrypting text — 101

Congratulations!

Well done! You have completed Chapter 6, Connect the world.

In this chapter you:

- ☑ explained how IP, DNS and HTTPS standards help to create a reliable and secure internet
- ☑ compared different wireless communication technologies and explained how they are used
- ☑ created your own encryption algorithm for a substitution cipher
- ☑ cracked an encryption algorithm by decoding a message and related your experience to modern internet security.

Key terms

Bit errors – When data gets changed during transmission

Bluetooth – Wireless standard for communicating between nearby devices

Crack – To find out how to get the plaintext from encrypted ciphertext without knowledge of the decryption algorithm or key

Decryption algorithm – Specific set of rules for turning a ciphertext message into a plaintext message, sometimes with the use of a key

Domain name – Human-readable label that is assigned to an IP address

Domain Name System (DNS) – Naming system and way of mapping from human-readable domain names to IP addresses

Encryption algorithm – Specific set of rules for turning a plaintext message into encrypted ciphertext, sometimes with the use of a key

Frequency analysis – Knowledge about how often each letter is used in text written in a particular language

Hypertext Transfer Protocol Secure (HTTPS) – Secure way of sending data between a web browser and a website

Key – Piece of information that can be used with a decryption algorithm to decrypt a ciphertext message

Packet loss – When 'packets' of data fail to reach their destination

Substitution cipher – Way of encoding messages where a letter is replaced by another letter or symbol

Top-level domain (TLD) – The last part of the domain. For example .com

4G and 5G – Versions of cellular network technology with improvements over previous generations

Reflect: Discuss: Did you enjoy cracking the codes? Explain your answer.

Workbook page 78: Complete the '**Reflection**' task.

Chapter 7 The power of data

Project: Help the school to make a student-led decision about where to spend some money

In this chapter, you will:

- create a spreadsheet to model a song royalties calculator
- make data-driven decisions using primary keys and conditional formatting
- design and use data capture forms.

End of chapter project: Data capture

Help the school to make a student-led decision about where to spend some money

Chapter 7.1 Modelling data to make decisions

What do we already know?

- A spreadsheet can perform calculations using operators and functions.
- Data is used to solve problems in industry.
- Computers can safely imitate dangerous situations, such as the machine learning asteroid intersect project that you saw in Chapter 1.

Key terms

Simulator – App or machine that realistically models real-world processes and conditions

Simulator games

A **simulator** is an app or machine that uses data to realistically model real-world processes and conditions.

Simulator games are a popular genre of video games. Some examples include:

- The Sims™
- RollerCoaster Tycoon®
- Microsoft Flight Simulator
- Football Manager™
- Chef Life

There are also some really niche simulator games such as a lawn mowing simulator and power wash simulator!

Discuss 1

Which simulator games have you seen or played?

Simulators in industry

Simulators are widely used in industry. Simulators can:

- safely and conveniently train people
  - a surgeon can practise operations
  - a pilot can practise landing a plane
- test process changes
  - a shop owner can test the impact of supply changes on customer orders
  - a construction worker can test the impact of building changes on health and safety procedures
- predict the future
  - an agricultural worker can see the potential destructive impact of storm paths
  - a salesperson can see current market trends on sales.

Although simulators are initially expensive, they can greatly reduce the cost of physically training people or testing changes to processes.

Workbook page 79: Complete Task A, '**Simulators in industry**'.

Discuss 2

What are the disadvantages of using a training simulator?

Simulators rely on data. The predictions and decisions made using a simulator are only as accurate as the data that they use. The data can come from:

- observations
- measurements
- sensors
- surveys
- experts
- historical records
- machine learning from simulator use.

Discuss 3

Why is accuracy of data important in simulations?

Project: Help the school to make a student-led decision about where to spend some money

A spreadsheet can be used to simulate a simple real-world scenario using data. The data and calculations can be easily changed to help to make informed data-driven decisions. Changes are carried throughout the model so that the impact can be analysed in detail before making decisions in the real world.

Making informed decisions

A singer has released a new song and two radio stations have agreed to play the song for a two-week trial. Each radio station wants exclusive access to the new song. Both radio stations have produced a schedule of guaranteed daily plays.

A royalty is a payment made to a creator each time their work is used. The first station will pay 0.30 royalties each time they play the song. The second radio station will pay 0.35 royalties but will play the song less frequently.

The singer wants to make enough money to go on holiday and has a target of 100. Help them to decide which radio station will get them to their target in the shortest time.

Radio station 1

Build 1:

Follow these steps:

1. Create a new spreadsheet and name it 'Royalties simulator'.
2. Right-click on the tab and rename it 'Station 1'.

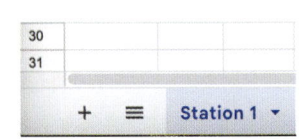

3. Enter the data.
4. Add a third column and name it 'Amount earned'. This column will use the calculation: royalty multiplied by times played. Type the formula '=$B$1*B4' into cell C4. This will work out the amount earned on Day 1. The result is 0.90.

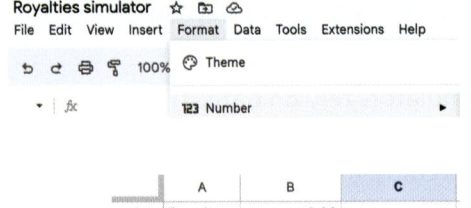

5. Click on cell C4 and drag the marker in the bottom-right corner down to cell C17. This will copy your formula down. In the formula, the cell 'B1' is written as '$B$1' so that the value from B1 stays the same when copied down. Cell 'B4' doesn't have any '$' so the value will automatically change to match the row.

Tip

You can format this into your local currency and change the decimal places from the formatting menu.

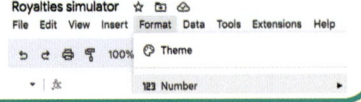

Chapter 7.1

6. Add a column D and name it 'Total amount'. In cell D4 type the formula '=C4' to give you the amount earned at the end of Day 1.

7. In cell D5 type the formula '=D4+C5'. This will give you the amount earned at the end of Day 2. Click on cell D5 and then drag the bottom-right marker down to D17 to work out the total amount earned throughout the trial.

8. Add two headers at the top of the sheet for 'Average earned' and 'Days until target'.

	A	B	C	D
1	Royalty	0.30	Average earned	
2	Target	100	Days until target	
3	Date	Times played	Amount earned	Total amount
4	Day 1	3	0.90	0.90
5	Day 2	5	1.50	2.40
6	Day 3	1	0.30	2.70
7	Day 4	5	1.50	4.20
8	Day 5	2	0.60	4.80
9	Day 6	1	0.30	5.10
10	Day 7	8	2.40	7.50
11	Day 8	5	1.50	9.00
12	Day 9	4	1.20	10.20
13	Day 10	1	0.30	10.50
14	Day 11	7	2.10	12.60
15	Day 12	2	0.60	13.20
16	Day 13	1	0.30	13.50
17	Day 14	4	1.20	14.70

9. In cell D1, work out the average number of plays and then multiply that by the royalty amount. '=AVERAGE(B4:B17)*B1'.

10. In cell D2, work out the target minus the guaranteed trial money from the first 14 days. Then divide the answer by the average earned. '=(B2-D17)/D1'.

Workbook page 79: Complete Task B Question 1, '**Data-driven decisions**'.

Tip
Format your simulator so that it is easy to see the important information.

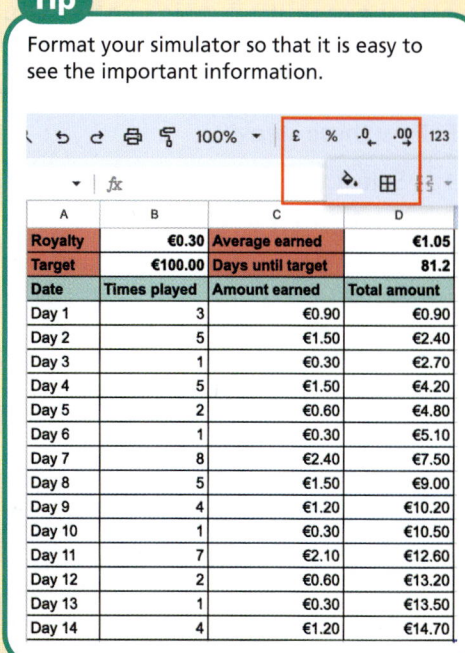

Project: Help the school to make a student-led decision about where to spend some money

Radio station 2

Build 2:

Follow these steps:

1. Right-click on the 'Station 1' tab and select 'Duplicate'. Rename the new tab 'Station 2'. Highlight the 'Times played' values and press 'Delete'.

	A	B	C	D
1	Royalty	€0.30	Average earned	#DIV/0!
2	Target	€100.00	Days until target	#DIV/0!
3	Date	Times played	Amount earned	Total amount
4	Day 1		€0.00	€0.00
5	Day 2		€0.00	€0.00
6	Day 3		€0.00	€0.00
7	Day 4		€0.00	€0.00
8	Day 5		€0.00	€0.00
9	Day 6		€0.00	€0.00
10	Day 7		€0.00	€0.00
11	Day 8		€0.00	€0.00
12	Day 9		€0.00	€0.00
13	Day 10		€0.00	€0.00
14	Day 11		€0.00	€0.00
15	Day 12		€0.00	€0.00
16	Day 13		€0.00	€0.00
17	Day 14		€0.00	€0.00

2. Change the royalty amount to '0.35' and type the number of guaranteed daily plays from station 2.

	A	B	C	D
1	Royalty	€0.35	Average earned	€0.83
2	Target	€100.00	Days until target	107.2
3	Date	Times played	Amount earned	Total amount
4	Day 1	3	€1.05	€1.05
5	Day 2	4	€1.40	€2.45
6	Day 3	1	€0.35	€2.80
7	Day 4	3	€1.05	€3.85
8	Day 5	2	€0.70	€4.55
9	Day 6	1	€0.35	€4.90
10	Day 7	3	€1.05	€5.95
11	Day 8	4	€1.40	€7.35
12	Day 9	2	€0.70	€8.05
13	Day 10	1	€0.35	€8.40
14	Day 11	3	€1.05	€9.45
15	Day 12	2	€0.70	€10.15
16	Day 13	1	€0.35	€10.50
17	Day 14	3	€1.05	€11.55

Tip

You will see errors in your spreadsheet until you add the new values for Station 2.

Workbook page 80: Complete Task B, '**Data-driven decisions**'.

Data visualisation

Modelling data in spreadsheets is a good way to see the data clearly and make decisions. You can also visualise the data using charts or export it to use in another visualisation program.

Build 3:

Follow these steps:

1. Select the 'Station 1' tab. Export the 'Station 1' tab to a csv file.

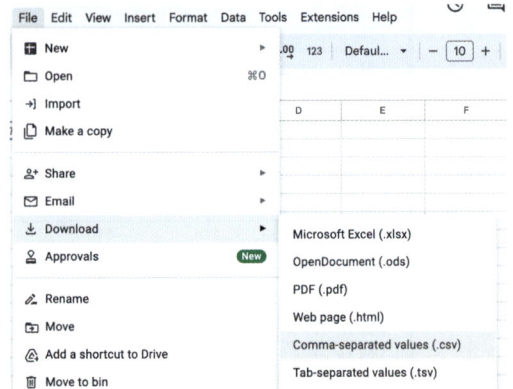

2. Your teacher will share the link to a Scratch project. Right-click the 'Plays' list on the Stage. Select 'import'.

108 Chapter 7 The power of data

Chapter 7.1

3. Navigate to your csv file and click 'Open'. Type '2' to import column 2 of your data.

4. The list has 3 summary items at the beginning that you do not need. Click on each item and then click the cross.

5. You will have a list of 14 items, which are the number of guaranteed plays each day.

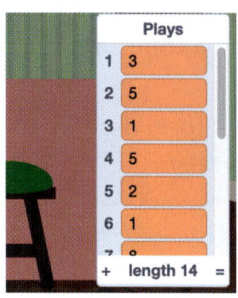

6. Click on the green flag. A column chart will be drawn at the bottom of the Stage. The columns visually represent the plays each day.

7. You will be asked to enter the royalty amount '0.40' and the target amount '100'.

8. Click on the singer to see how many days until they can go on holiday.

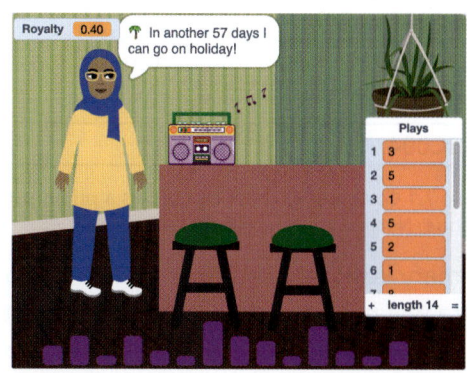

As with the spreadsheet model, you can change the royalty or the target to see what effect this has. You can also clear the 'Plays' list and import different data.

With this method of visualisation you can customise the costumes, backdrop, speech bubbles and chart colour to make a visualisation that is personal to you.

Reflect: Discuss: Did you prefer seeing the data in the spreadsheet or represented graphically in the Scratch project? Why?

Workbook page 80: Complete the '**Reflection**' task.

Project: Help the school to make a student-led decision about where to spend some money

Chapter 7.2 Organising and formatting data

What do we already know?

- Data is used to solve problems.
- A spreadsheet can perform calculations using operators and functions.
- Data in spreadsheets, word processors and presentation software can be formatted using colour.
- A database is a structured collection of data that can be searched and updated.
- A data capture form is a form designed to collect data from people, often digitally.

Key terms

Record – Row in a database table

Field – Column in a database table

Primary key – Assigned field in a database table that contains data that is both unique and not empty for every record

Conditional formatting – Rule with criteria that, if met, changes the appearance of cells within a spreadsheet

Organising data

A database is a structured collection of data that can be searched and updated. Databases use tables to store information. Each row in a table is called a **record**. Each column in a database table is called a **field**. Every record is made up of a collection of fields.

A **primary key** is a field in a database table that contains data that is unique for every record. A primary key cannot be empty in any record.

Customer ID ✓	Name ✗	Nearest store ✗
ABC001	Priya	Main Street
ABC002	Haroon	
ABC003	Raju	High Street
ABC004	Priya	Third Street

Table 7.1 Customer ID could be used as a primary key. Name has data that is not unique. Nearest store has data that is empty.

Primary keys help you to organise, find and link data. Sometimes data will not have any fields that are both unique and not empty. In these situations a new field can be created.

Name ✗	Industry ✗	Job role ✗
Priya	Health	Pharmacist
Haroon		
Raju	Health	Optician
Priya	Manufacturing	

Table 7.2 None of the fields are valid as a primary key.

Chapter 7.2

ID ✓	Name ✗	Industry ✗	Job role ✗
1	Priya	Health	Pharmacist
2	Haroon		
3	Raju	Health	Optician
4	Priya	Manufacturing	

Table 7.3 ID is a new field that is valid as a primary key.

> **Discuss 4**
>
> In your royalty model from the last lesson, which field would make the best primary key? Why?
>
> | Date | Times played | Amount earned | Total amount |
> |---|---|---|---|
> | Day 1 | 3 | 1.20 | 1.20 |
> | Day 2 | 5 | 2.00 | 3.20 |

Workbook page 81: Complete Task A, '**Identify primary keys**'.

Conditional formatting

Conditional formatting is a useful feature for visually highlighting data in spreadsheets. Rules are used to automatically change the appearance of cells.

Day	Max temperature
Monday	18
Tuesday	21
Wednesday	20
Thursday	23

In this example, the rule is that if a cell value in column B is greater than 20 it should have bold, white text and a red background.

Conditional formatting can also be used to add a colour scale to values. All cells are given a colour based on their position on the colour scale.

Day	Max temperature
Monday	18
Tuesday	21
Wednesday	20
Thursday	23

In this example, the highest value is red, the middle value is orange and the lowest value is green.

Project: Help the school to make a student-led decision about where to spend some money

Visualising informed decisions

The singer from the last lesson has been sent a DJ schedule from radio station 1. They want to see how often each DJ is playing their song. You are going to help them by creating a visual representation of the data.

Linking tables

Build 4:

Follow these steps:

1. Open your 'Royalties simulator' from the last lesson and create a new tab called 'DJ Schedule'. Add the following data:

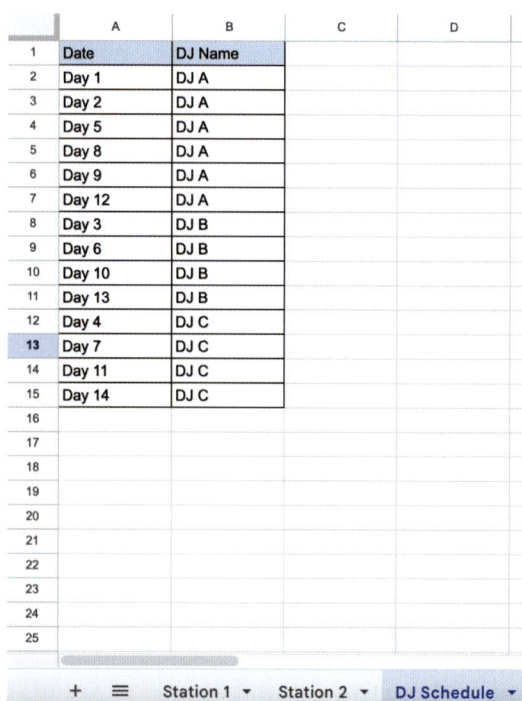

2. Spreadsheets use primary keys to link data between tabs and tables. Your 'Station 1' table and your 'DJ Schedule' table both have the field 'Date'. Date contains unique values and is always populated. Click on the 'Station 1' tab. Right-click on the column A heading and select 'Insert 1 column to the left'. Call your new column 'DJ' and format your table.

A	B	C	D	E
	Royalty	€0.40	Average earned	€1.40
	Target	€100.00	Days until target	57.4
DJ	Date	Times played	Amount earned	Total amount
	Day 1	3	€1.20	€1.20
	Day 2	5	€2.00	€3.20
	Day 3	1	€0.40	€3.60
	Day 4	5	€2.00	€5.60
	Day 5	2	€0.80	€6.40
	Day 6	1	€0.40	€6.80
	Day 7	8	€3.20	€10.00
	Day 8	5	€2.00	€12.00
	Day 9	4	€1.60	€13.60
	Day 10	1	€0.40	€14.00
	Day 11	7	€2.80	€16.80
	Day 12	2	€0.80	€17.60
	Day 13	1	€0.40	€18.00
	Day 14	4	€1.60	€19.60

Chapter 7.2

3. A 'vlookup' formula is used to link tables together. Add this formula to cell A4 '=vlookup(B4,'DJ Schedule'!$A$2:$B$15,2,FALSE)'.

	A	B	C	D	E
1		Royalty	€0.40	Average earned	€1.40
2		Target	€100.00	Days until target	57.4
3	DJ	Date	Times played	Amount earned	Total amount
4	DJ A	Day 1	3	€1.20	€1.20
5		Day 2	5	€2.00	€3.20

A specific 'Date' from the 'Station 1' table.

The 'DJ' column in the 'DJ Schedule' table

=vlookup(B4, 'DJ Schedule'!$A$2 : $B$15,2,FALSE)

The full table on the 'DJ schedule' tab

Return only an exact match

4. Drag the bottom-right marker down to fill the whole list with the formula.

	Royalty	€0.40	Average earned	€1.40
	Target	€100.00	Days until target	57.4
DJ	Date	Times played	Amount earned	Total amount
DJ A	Day 1	3	€1.20	€1.20
DJ A	Day 2	5	€2.00	€3.20
DJ B	Day 3	1	€0.40	€3.60
DJ C	Day 4	5	€2.00	€5.60
DJ A	Day 5	2	€0.80	€6.40
DJ B	Day 6	1	€0.40	€6.80
DJ C	Day 7	8	€3.20	€10.00
DJ A	Day 8	5	€2.00	€12.00
DJ A	Day 9	4	€1.60	€13.60
DJ B	Day 10	1	€0.40	€14.00
DJ C	Day 11	7	€2.80	€16.80
DJ A	Day 12	2	€0.80	€17.60
DJ B	Day 13	1	€0.40	€18.00
DJ C	Day 14	4	€1.60	€19.60

Spot the trend

Build 5:

Follow these steps:

1. Highlight the values in the 'Times played' column. Then go to 'Format' and select 'Conditional formatting'.

	A	B	C	D	E
1		Royalty	€0.40	Average earned	€1.40
2		Target	€100.00	Days until target	57.4
3	DJ	Date	Times played	Amount earned	Total amount
4	DJ A	Day 1	3	€1.20	€1.20
5	DJ A	Day 2	5	€2.00	€3.20
6	DJ B	Day 3	1	€0.40	€3.60
7	DJ C	Day 4	5	€2.00	€5.60
8	DJ A	Day 5	2	€0.80	€6.40
9	DJ B	Day 6	1	€0.40	€6.80
10	DJ C	Day 7	8	€3.20	€10.00
11	DJ A	Day 8	5	€2.00	€12.00
12	DJ A	Day 9	4	€1.60	€13.60
13	DJ B	Day 10	1	€0.40	€14.00
14	DJ C	Day 11	7	€2.80	€16.80
15	DJ A	Day 12	2	€0.80	€17.60
16	DJ B	Day 13	1	€0.40	€18.00
17	DJ C	Day 14	4	€1.60	€19.60

Project: Help the school to make a student-led decision about where to spend some money

2. Click on 'Colour scale', choose your scale colours and then click 'Done'.

Discuss 5
Did you find a relationship between the DJ and the times played?

Discuss 6
Why was linking tables and applying conditional formatting useful for visualising the relationship?

Workbook page 82: Complete Task B, '**Highlight scores**'.

Reflect: Is highlighting the entire row useful or not? Why do you think this?

Workbook page 82: Complete the '**Reflection**' task.

Chapter 7.3   Search and capture data

What do we already know?

- A database is a collection of data that has been organised in tables.
- Data can be filtered to results that match a keyword.
- Data can be captured through many sources, for example, sensors, connected devices and forms.

Key terms

Database query – Search tool that returns records that match the specified criteria

Data capture form – Form designed to collect data from people, often digitally

Searching data

Data stored in a spreadsheet can be searched to find only those rows that match specified criteria.

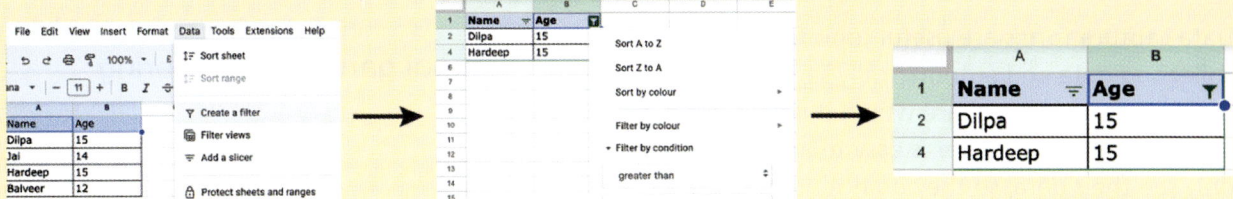

Data stored in a database is organised and easy to search. A database search is called a **database query**. Queries return data that match their criteria.

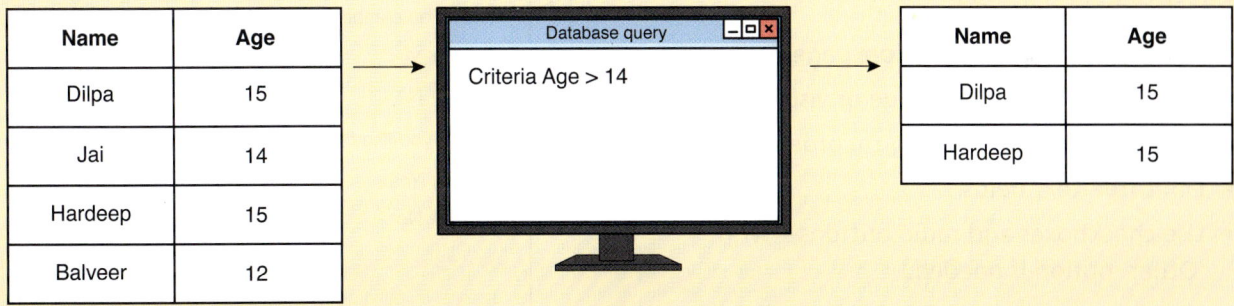

Workbook page 83: Complete Task A, '**Searching databases**'.

Capturing data

There are many ways to enter data into a database or a spreadsheet. Data can be typed manually, as you did earlier in this chapter, or entered automatically from sensor readings or from a **data capture form**.

Project: Help the school to make a student-led decision about where to spend some money

Data capture forms have a wide range of uses:

- website contact forms
- opinion survey, questionnaire or research forms
- registration or subscription sign-up forms
- customer order forms
- job application forms
- hospital patient forms
- ticket reservation forms for events
- customer satisfaction forms
- complaint or evaluation forms.

> **Discuss 7**
>
> Have you ever filled in a data capture form? If so, when?

Data capture forms are useful tools for collecting data and automatically sending it to a spreadsheet or a database where it is organised in a structured manner that is easy to analyse. Users only see their data, so they do not need to learn how to use a spreadsheet or database app.

However, not all forms are the same! Some forms are easier, more appealing and effective to use than others. Here are some tips for creating effective data capture forms:

- Keep it short, or use multiple page forms.
- Show a progress percentage or visual graphic
- Arrange questions in a logical order.
- Limit free-text fields.
- Use checkboxes and radio buttons – with 'Other' option if needed.
- Use expected answer tips or descriptions.
- Add a confirmation message when the form has been submitted.
- Use skip to section/question functionality based on other answers.

School garden volunteering

Progress 30%

- Which lunchtime can you come?
  Choose one day only
  ○ Monday   ◉ Wednesday
- Which activities are you able to help with?
  Choose as many as you like
  ☐ Litter collection ☐ Pulling weeds
  ☐ Painting
- Have you completed first aid training?
  ○ Yes   if yes, skip next question
  ○ No
- Would you like to do first aid training?
  ○ Yes
  ○ No

Stay safe

Make sure that you use an online form tool from an organisation with a good reputation for security. Comply with local laws and advice about storing and using data, especially when using sensitive or personal information. Avoid unwanted responses by restricting access to your form and using data validation rules for free-text fields.

Workbook page 83: Complete Task B, '**Effective forms**'.

Your teacher

Your teacher will give you a link to a form that they have created. Answer the questions and then submit the form.

Create a data capture form

Chapter 7.3

Build 6:

Your teacher will give you a link to the Forms app.

1. Create a new form.

   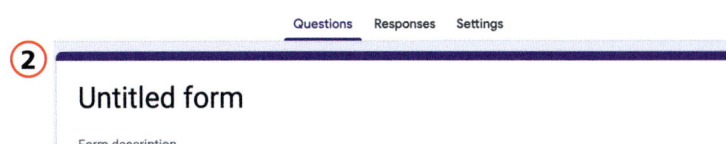

2. Add a form title and description.

3. Add your first question. Set up the correct answer type and add your answer options.

   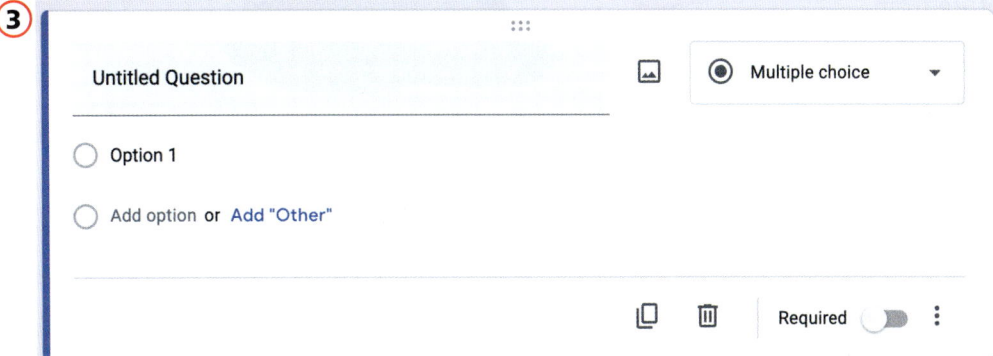

4. Choose whether an answer is required and add a description to help users to answer your question.

5. Add a second question.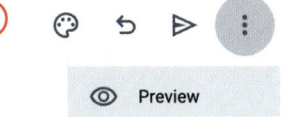

6. Experiment with the settings of your form to set the appearance and question functionality.

7. Click 'Preview' to see your form from a user's viewpoint.

   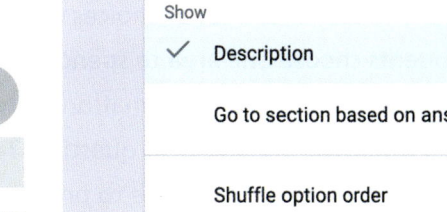

8. Complete your form to test it and then look at your answers on the Responses tab.

9. Click 'Link to Sheets', and send your response to a new spreadsheet.

Reflect: Which tips did you use when creating your data capture form?

Workbook page 84: Complete the '**Reflection**' task.

Project: Help the school to make a student-led decision about where to spend some money

Chapter 7.4   Ask your audience

What do we already know?

- Data capture forms are used to gather data.
- Captured data can be automatically sent to a spreadsheet.
- Effective forms are those that are appealing to use and capture structured data that can be analysed.

Project brief

Help the school to make a student-led decision about where to spend some money

You need to make a student-led decision based on data that you capture from students at your school.

You will:

- design a form to gather the opinions of students
- create and test your form
- survey students to capture their data
- use conditional formatting to help analyse your data
- present your process and decision to your class
- evaluate the effectiveness of your form.

Design your data capture form

Think about the information you need to collect. Your objective is to find out where students would like to spend the available money.

- What questions will you ask?
- Will you give students a set of choices or ask them to make suggestions?
- Will students choose one area to spend all of the money or divide it amongst several areas?
- Do you need any personal information such as age or class?
- In which order will you ask your questions?
- Which form features will you use to provide the best user experience?

Workbook page 85: Complete Task A, '**Design your form**'.

Use your online forms app to create the form using your Workbook plan as a guide.

Reflect: Did you have any challenges designing your form? If so, describe them.

Tip
Make sure that you link your form to a spreadsheet so that you can analyse your data in the next lesson.

Workbook page 85: Complete the '**Reflection**' task.

Chapter 7.5 Analyse your data

What do we already know?

- Data capture forms can link to a spreadsheet to store the data.
- Spreadsheets use formulae to calculate or restrict data.
- Spreadsheets can be formatted to help to analyse data and make data-driven decisions.

In the last lesson, you surveyed students using your data capture form. The data will be stored in your linked spreadsheet and is now ready to analyse.

You will use your knowledge of conditional formatting to help you to make and explain your student-led decision. You can use single colour or colour scale conditional formatting, and highlight single cells or whole rows.

Conditional formatting examples

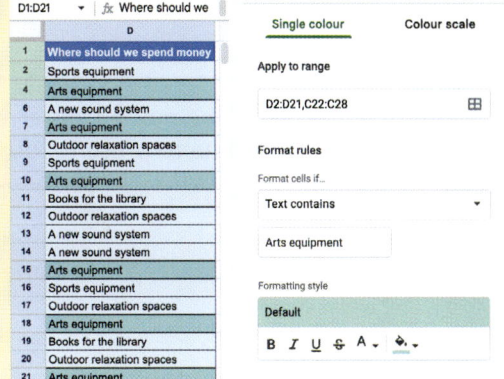

In the first example (top right), the students decided to spend the money on art equipment because that was the most popular choice.

In the second example (middle right), the students decided to spend 250 on musical instruments because that was the average amount suggested.

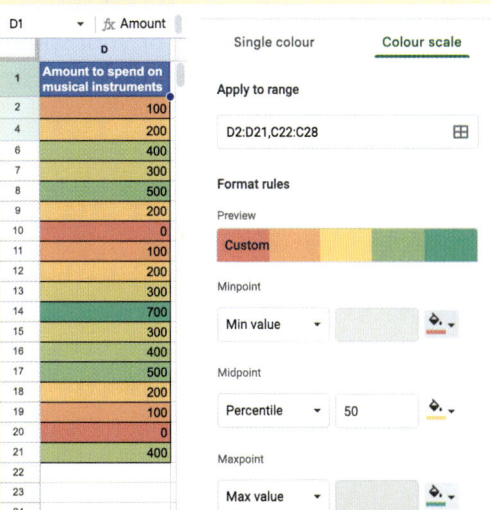

In the third example (bottom right), the students surveyed the whole school and spotted a trend. Younger students wanted to spend the money on sports but older students wanted to spend the money on computers. They decided to spend half of the money in each area.

> **Workbook** page 86: Complete Task A, '**Create your model**'.

In the next lesson, you will present your spreadsheet and your student-led decisions.

> **Workbook** pages 86 and 87: Complete Task B, '**Student-led decisions**'.

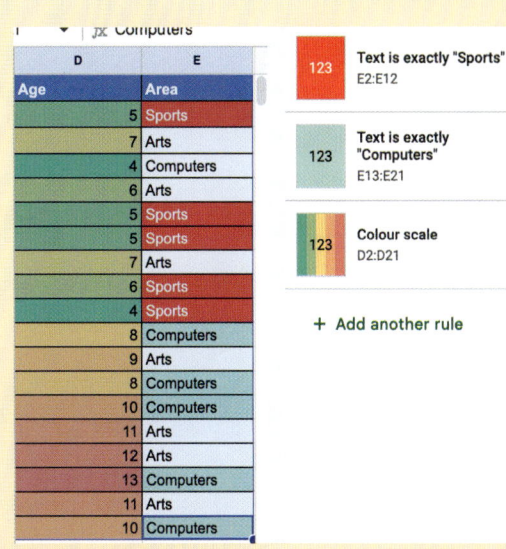

> **Reflect:** Share the challenges that you had while creating your model.

> **Workbook** page 87: Complete the '**Reflection**' task.

Project: Help the school to make a student-led decision about where to spend some money

Chapter 7.6 Present your decision

What do we already know?

- Data capture forms can be used to gather opinions.
- Data capture forms can link to spreadsheets.
- Conditional formatting helps to make data-driven decisions.

Showcase

Present your project. The money has arrived in the school bank account, so now it's time to present your findings.

Tip

Use the conditional formatting in your spreadsheet to visually back up your decision.

Workbook page 88: Complete task A, '**Reflection**'.

Discuss 8

What improvements did you identify that you would like to make to either your form or your spreadsheet?

Congratulations!

Well done! You have completed Chapter 7, 'The power of data'.

In this chapter you:

- ☑ created a spreadsheet to model a song royalties calculator
- ☑ made data-driven decisions using primary keys and conditional formatting
- ☑ designed and used data capture forms.

Key terms

Conditional formatting – Rule with criteria that, if met, changes the appearance of cells within a spreadsheet

Database query – Search tool that returns records that match the specified criteria

Data capture form – Form designed to collect data from people, often digitally

Field – Column in a database table

Primary key – Assigned field in a database table that contains data that is both unique and not empty for every record

Record – Row in a database table

Simulator – App or machine that realistically models real-world processes and conditions

Reflect: Why are spreadsheets useful tools for making data-driven decisions?

Workbook page 88: Complete the '**Reflection**' task.

Chapter 8: Create with code 3

Project: Plan, design, test and build a device to welcome visitors to your classroom

In this chapter, you will:

- learn about project planning in software development
- discover ways to provide a great user experience
- prototype ideas
- design and build a device and test it using a test plan
- showcase your designs to the class.

End of chapter project: Plan, design, test and build

Plan, design, test and build a device to welcome visitors to your classroom

Chapter 8.1 Project plans

What do we already know?

- Project plans can be really useful for making sure that a project is successful.
- The end user is the person who uses a project for its intended purpose.
- The BBC micro:bit is a tiny computer that has inputs and outputs that can be programmed to make fun projects.

Key terms

Project plan – Carefully planned tasks that ensure a project meets a deadline to the correct requirements

Software development – Creating software (programs) for a specific user requirement or need

Scope creep – Features that were not in the original requirements getting added over time

User-centered design – Involving the end user of a product in every stage of the design process

Discuss 1

What projects have you used a BBC micro:bit for? What inputs and outputs did you use?

Projects need a plan

To achieve a goal for a project, it is important to have a clear plan for achieving that goal. This is called a **project plan**.

Without a plan, a project might take longer to achieve and it may not meet the requirements.

Discuss 2

Can you think of a time when you have tried to complete a project without a plan? Was it still successful? Would a plan have helped you?

Project: Plan, design, test and build a device to welcome visitors to your classroom

Project plans in software development

When an organisation needs to create a new piece of software, a project plan is created to ensure that the project:

- meets the deadlines
- meets the needs of the end user
- doesn't go over budget.

A project plan for **software development** might include the following:

- The key objectives – What should be achieved from the piece of software?
- Success criteria – How will you know that the project is a success?
- Stakeholders – Who are the key people involved in ensuring that the project is a success?
- Scope – What are the defined functions or features of the software?
- Budget – How much money should the project cost?
- Dependencies – Which parts of the project depend on other parts being finished first?
- Timeline – How long will it take to complete each part of the project and when should the whole project be finished?

Workbook page 89: Complete Task A, '**Order the tasks**'.

Case study: The new computer game

An author created a book series that was popular with children all over the world. They wanted to expand their book series into a computer game where players could follow the story as if they were any of the main characters in the book. It was August and they wanted to release the game in 18 months' time, when their final book was to go on sale.

They worked with a game development company who told them that 18 months was an unrealistic time to create a high-quality game. The author wanted to go ahead with the project, despite being warned that the deadline may not be met. In order to meet the deadline, the project was not tested correctly. On the launch date, players were disappointed that their characters would randomly get stuck in corners of a room, which made them have to quit the game. The customers were very unhappy and demanded refunds.

Discuss 3

There were two main factors that made this project fail. What were they?

The main reasons that projects fail are:

- **Scope creep** – Additional functions and features are added during the development process that weren't originally planned or budgeted for.
- **Missing stakeholders** – People are left out of important discussions so the full range of experience and skills are not used for the project.
- **Dependencies not factored in** – If a task relies on another task being completed first, then this must be properly planned into the project.

Workbook page 90: Complete Task B, '**Case study: Mobile ticketing**'.

User-centered design

An important stakeholder in software development is the end user. You must consider how they will use the software, whether it meets their needs and how easy it is for them to use it. **User-centered design** involves the end user at every stage of development.

Fast food order screens

Imagine a fast food restaurant wants to introduce large screens that allow the customer to select meals and pay without speaking to a server.

When developing a change to an ordering system such as this, the customer (end user) must be involved or the restaurant could lose sales. Here are some examples of how a user could be involved.

- Interviews – Ask customers how they typically decide what they would like to order and what method they usually use to pay.
- Observations – Observe typical orders using the current system, then provide draft versions of the new system and observe how easily customers use it to place an order.
- Surveys – After they have used the new system, ask customers to complete a feedback survey to find out their opinions of it.

After interviewing, observing and surveying customers the software developers realised that many customers who used the restaurant had English as an additional language. They requested that a translation feature could be added to help them to order in their first language, so as to make the process faster and easier and lead to increased sales.

Workbook pages 90 and 91: Complete Task C, '**Effective design**'.

Reflect: Share the key things you learnt from this lesson. What did you find most interesting?

Workbook page 91: Complete the '**Reflection**' task.

Chapter 8.2   Test plans

What do we already know?

- Programs can contain errors that require identifying and fixing. This is called debugging.
- EduBlocks can be used to write Python code by dragging and dropping blocks.

Key terms

Iterative testing – Running your program as you develop it to check that there are no syntax (or block) errors

Test plan – List of tests to be carried out on a finished program to ensure that it works as expected

Test plan

When you are creating a new program, it is important to have a **test plan**. A test plan allows you to think about all of the things a user might do with your program and whether those things might produce an error. It also allows you to plan how you will make sure that your program meets its requirements.

Here are the first three tests from a test plan that a student has made for an insect quiz for the BBC micro:bit. The quiz should:

- have a welcome message
- ask a question
- allow the user to enter the answer by pressing an input button to increase a number.

#	Description	Input	Expected output	Result	Fix (if needed)
1	A welcome message appears	Switch on / play	'Welcome to my game. Press A to start'	As expected	Pass
2	User presses A to start	A	Question 1: How many legs does an insect have?	Nothing happened	The input was set to B instead of A, this has been fixed
3	The number on the screen increases by 1 each time B is pressed	B, 6 times	The number increases to 6 when B is pressed 6 times	The number reached 7 after 6 presses	The count began at 1 instead of 0

By following the test plan, the student identified and fixed two errors.

Project: Plan, design, test and build a device to welcome visitors to your classroom

Micro:bit in EduBlocks

You have used EduBlocks before to write Python code. You have also used a micro:bit before to create physical computing projects with MakeCode. EduBlocks allows you to use a micro:bit simulator to write and test projects for the micro:bit.

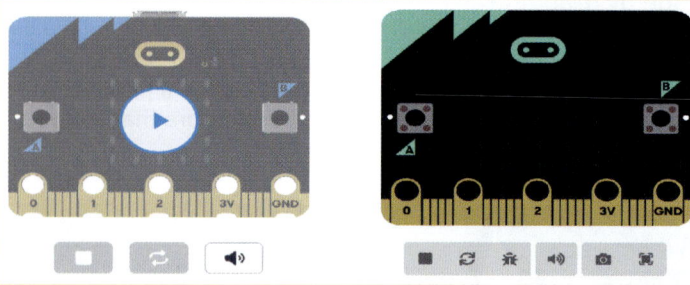

EduBlocks simulator   MakeCode simulator

In this chapter you will be writing Python code in EduBlocks for the micro:bit.

Workbook page 92: Complete Task A, '**Follow a test plan**'.

Discuss 4

Which errors did you spot in the code? How did you fix them?

Workbook page 93: Complete Task B, '**Write and follow a test plan**'.

Discuss 5

Which errors did you spot in the code? How did you fix them? Did you miss any tests?

Reflect: Why is testing important?

Workbook page 93: Complete the '**Reflection**' task.

Chapter 8.3 Prototypes

What do we already know?

- Testing products with end users throughout the design process improves the usability of the final product.

Key terms

Prototype – Sketch or model to test a concept for a project

Prototypes

Prototypes are used in software development to see whether a project is possible and to identify problems that may occur when a user begins to use the software.

Prototypes often begin as sketches with some key annotations. These can then be shown to potential users who are asked to give their feedback.

Example prototype sketch

A software development company has been hired to create a new feature for a music streaming application. The music streaming service would like the feature to make it really easy for a user to play their favourite songs with one touch of a button. There should be a recognisable icon from the main screen and the playlist should play automatically.

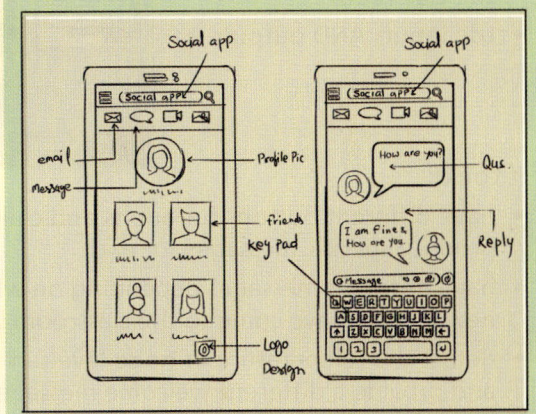

Here are the prototype sketches that have been provided by the software development company.

Discuss 6

Do you think that the sketches show the key requirements for the project? What could be improved?

Creating prototype sketches for a project means that you can ask for feedback in the early stages of development to see what might need to be improved.

Workbook page 94: Complete Task A, '**Does it meet the requirements?**'.

Project brief

Plan, design, test and build a device to welcome visitors to your classroom

Your device must:
- use a BBC micro:bit
- welcome visitors to the classroom
- use inputs AND outputs.

It could:
- have different buttons for adults and children to press that produce specific messages for each
- have different messages depending on whether the visitor is new or they have come to the classroom before
- detect the temperature or noise levels in the classroom and provide a different welcome message for different temperatures or volumes of noise
- have a secret message option when both A and B are pressed at the same time.

Workbook pages 95 and 96: Complete Task B, '**Your prototype sketch**', and Task C, '**Your improved prototype sketch**'.

Reflect: Why is prototyping important when creating a new product?

Workbook page 96: Complete the '**Reflection**' task.

Create your device

What do we already know?

- EduBlocks can be used to write code for a micro:bit.
- Taking time to plan your project leads to success.

Project brief

Plan, design, test and build a device to welcome visitors to your classroom

Student examples

1. Almaz created a musical device because music is very welcoming. Each day people can decide which button to press and then get a surprise tune. Using funny tunes adds humour and creates a talking point.

2. Niru created this visual welcoming device. People can chose to turn the device on if they are happy, enjoy making things light up and want to get involved. Alternatively, if they are annoyed or sad, they may choose to turn the welcoming device off and not take part.

3. Ajeet created this silly welcoming device that hangs from a string and uses different types of input and output. The device reacts to shaking and button presses by displaying an image and playing a sound. It is intended to be a fun way to welcome people into the room.

Build 1:

Remind yourself of your prototype from last lesson and then use EduBlocks to create the code for your micro:bit device.

There is space in your Workbook to plan your code and make a note of any design choices such as names of images or music that you want to include in your code.

Tip

Build your code gradually for one input and output at a time. Use the Simulator to test each part before combining and testing the device as a whole.

Workbook page 97: Complete Task A, '**Plan your build**'.

Reflect: How successful were you at sticking to your prototype without scope creep?

Workbook page 97: Complete the '**Reflection**' task.

Test and refine your device

What do we already know?

- Programs can contain errors that require identifying and fixing. This is called debugging.
- Successful projects use test plans.
- Testing products with end users throughout the design process improves the usability of the final product.

Project brief

Plan, design, test and build a device to welcome visitors to your classroom

Your device must:

- welcome visitors to the classroom
- use inputs <u>and</u> outputs.

Test your device

Testing whether your device meets the brief is an essential part of a successful project.

Workbook page 98: Complete Task A, '**My test plan**'.

Build 2:

Use your test plan results and fix information to make any necessary changes to your device.

Download your project to your physical micro:bit device.

You should also test your project with end users and use your observations and their feedback to improve your device.

Workbook pages 98 and 99: Complete Task B, '**User research**'.

Build 3:

Use your observations and end-user feedback to make any necessary changes to your device.

Reflect: What differences did you notice between testing your device in the simulator and testing your physical device with end users?

Workbook page 99: Complete the '**Reflection**' task.

Showcase your device

What do we already know?

- A showcase is a time to share what you have created with an audience.
- It is a good idea to practise what you will say for your showcase by talking through it aloud.

What does it mean to showcase your project?

Showcasing is a way of sharing your project with other people.

Tips for showcasing your project:

- Speak loudly and clearly.
- Make sure you stand so that your audience can clearly see your device.
- Give the audience an opportunity to press the different inputs so that they can see the different outputs.

Showcase

Showcase your 'classroom welcome' device. You should:
- state how your device meets the requirements
- give your audience a chance to use your device
- describe how you tested your project to make sure that it didn't contain errors.

Workbook page 100: Complete Task A, '**Reflection**'.

Congratulations!

Well done! You have completed Chapter 8, 'Create with code 3'.

In this chapter you:

- ☑ learnt about project planning in software development
- ☑ discovered ways to provide a great user experience
- ☑ prototyped ideas
- ☑ designed and built a device and tested it using a test plan
- ☑ showcased your design to the class.

Key terms

Project plan – Carefully planned tasks that ensure a project meets a deadline to the correct requirements

Prototype – Sketch or model to test a concept for a project

Scope creep – Features that were not in the original requirements getting added over time

Software development – Creating software (programs) for a specific user requirement or need

Test plan – List of tests to be carried out on a finished program to ensure that it works as expected

User-centered design – Involving the end user of a product in every stage of the design process

Reflect: How well did your device create a welcoming environment for visitors to your classroom?

Workbook page 100: Complete the '**Reflection**' task.

Glossary of key terms

Analogue sound – Sounds that are all around us in the nondigital world

Application software – Software used for a specific task, such as a word processor

Argument – Input to a subroutine call

Artificial general intelligence – Future computer systems that might be able to perform any task in a human-like way

Automation – Carrying out tasks with little to no human interaction

Bit errors – When data gets changed during transmission

Bluetooth – Wireless standard for communicating between nearby devices

Boolean expression – Expression that either has a TRUE or FALSE outcome

Boolean logic – Form of algebra that uses the operators AND, OR and NOT

Call (a subroutine) – To tell the program to execute the code within a subroutine

Camera direction – Shots and angles used to describe the position, view and movement of a camera

Characters – Numbers, letters and punctuation, for example, A, " or 9

Citation – Credit given to identify the original author of a piece of work

Cloud storage – Space to save, access and manage files that is accessed over the internet

Colour depth – Number of bits (1s and 0s) that are available for each pixel. The higher the number of bits, the more colours that are available

Community guidelines – Behaviours that are expected from members of an online community

Conditional formatting – Rule with criteria that, if met, changes the appearance of cells within a spreadsheet

Cookie – Small piece of information that a website stores on your computer

Coworking buildings – Workspaces where people go to work together with others in the same physical location

Crack – To find out how to get the plaintext from encrypted ciphertext without knowledge of the decryption algorithm or key

Cyberbullying – Bullying someone using online tools

Database query – Search tool that returns records that match the specified criteria

Data capture form – Form designed to collect data from people, often digitally

Data type – Classification such as string or integer that tells the computer how to work with a value

Decimal – Base-10 number system that uses 10 digits, from 0 to 9

Decryption algorithm – Specific set of rules for turning a ciphertext message into a plaintext message, sometimes with the use of a key

Denary – Alternative for 'decimal'

Domain name – Human readable label that is assigned to an IP address

Domain Name System (DNS) – Naming system and way of mapping from human-readable domain names to IP addresses

Encryption algorithm – Specific set of rules for turning a plaintext message into encrypted ciphertext, sometimes with the use of a key

Fair use – Exceptions that allow the use of work that is under copyright

Field – Column in a database table

Frequency analysis – Knowledge about how often each letter is used in text written in a particular language

Function (Python) – Type of subroutine that returns a value. Python has built-in functions that are premade for you to use. You can also write your own functions in Python

Future technologist – Expert who uses current technology trends to make predictions about the future

Hypertext Transfer Protocol Secure (HTTPS) – Secure way of sending data between a web browser and a website

Image recognition – Identification of objects in images by computer systems using AI

Integer – Whole number

Key – Piece of information that can be used with a decryption algorithm to decrypt a ciphertext message

Logic gates – Used in logic circuits in computers. They are based upon the Boolean logic operators AND, OR and NOT

Machine learning – Type of AI that uses data and algorithms to gradually produce more accurate results in a way that imitates how humans learn

Machine learning model – Algorithm plus data that can be used to make decisions or predictions

News script – Text that details dialogue, timing and camera actions of a news item

Online collaboration – People working together in apps accessed over the internet

Operating system – Software that manages the hardware, software and resources of a computer

Packet loss – When 'packets' of data fail to reach their destination

Parameter – Special named variable that is part of a subroutine definition

Permission – When a website or app asks to access your data or sensors

Pixel – Block of colour that forms part of an image. A binary code is used to represent each colour

Plagiarism – Copying someone else's work then saying it is a new piece of work

Policy – Document explaining what you can expect from a website or service, including how it will use your data

Primary key – Assigned field in a database table that contains data that is both unique and not empty for every record

Project plan – Carefully planned tasks that ensure a project meets a deadline to the correct requirements

Prototype – Sketch or model to test a concept for a project

Python – Text-based programming language that is popular in industry

Recommendation system – Computer system that makes predictions on content that a user would like, using AI

Record – Row in a database table

Scope creep – Features that were not in the original requirements getting added over time

Simulator – App or machine that realistically models real-world processes and conditions

Software development – Creating software (programs) for a specific user requirement or need

Statement – Complete instruction that the computer can run

Storage capacity – The maximum amount of data that can be stored

Subroutine – Named section of code that can be called and performs a specific task

Substitution cipher – Way of encoding messages where a letter is replaced by another letter or symbol

Supervised machine learning – Type of machine learning where humans label data to help the machine learning model classify new data

System software – Software that controls how the computer operates

Test plan – List of tests to be carried out on a finished program to ensure that it works as expected

Text-based programming language – Allows you to type text instructions that a computer can run

Top-level domain (TLD) – The last part of the domain. For example .com

Training data – Data used to train a machine learning model

Trolling – Deliberately unhelpful behaviour in an online community

Unsupervised machine learning – Type of machine learning where the AI detects patterns in data and learns how to classify new data

User-centered design – Involving the end user of a product in every stage of the design process

Utility software – Software that provides additional support for a system, such as antivirus or disk cleanup management

Voice assistant – AI system that can perform useful actions in response to voice commands

Wellbeing – The way people feel about themselves and their lives

4G and 5G – Versions of cellular network technology with improvements over previous generations

Acknowledgements

Screenshots

Support materials and screenshots are licensed under the Creative Commons Attribution-ShareAlike 2.0 license. We are grateful to the following for permission to reproduce screenshots. In some instances, we have been unable to trace the owners of copyright material, and we would appreciate any information that would enable us to do so.

Python Software Foundation: Permission obtained to use screenshots demonstrating Python programming language features.

Scratch Foundation: Authorised usage of screenshots showcasing Scratch programming environment elements.

Machine Learning for Kids: Kindly granted permission for the incorporation of screenshots illustrating machine learning concepts and tools designed for educational purposes.

EduBlocks: Granted clearance for the inclusion of screenshots depicting EduBlocks interface and functionalities.

Scratch is developed by the Lifelong Kindergarten Group at the MIT Media Lab: p.1, p.10, p.12–19, p.42, p.46, p.74, p.96, p.108–109.

PSF's License Agreement and PSF's notice of copyright, i.e., "Copyright (c) 2001 Python Software Foundation; All Rights Reserved" are retained in Python 2.0.1 alone or in any derivative version prepared by Collins (Licensee): p.65–69, p.71–78, p.80.

EduBlocks: p.67–68, p.74, p.76.

Images

We are grateful for the following for permission to reproduce their images:

p.8 Jittawit21/ Shutterstock, p.8 VesnaArt/Shutterstock, p.10 ESB Professional/Shutterstock, p.51 A_B_C/Shutterstock, p.66 metamorworks/Shutterstock, p.91 jamesteohart/Shutterstock, p.99 PeopleImages.com - Yuri A/Shutterstock, p.99 Hilmi Abedillah/Shutterstock, p.104 boommaval/Shutterstock, p.104 WeAre/Shutterstock, p.104 Q88/Shutterstock, p.126 Scharfsinn/Shutterstock.